THE MANUAL OF
PRACTICAL HOMESTEADING

The Manual
of Practical Homesteading

by John Vivian

Rodale Press Book Division
Emmaus, Pa.

10 9 hardcover

10 9 paperback

PRINTED IN THE UNITED STATES OF AMERICA
on recycled paper.

Library of Congress Cataloging in Publication Data

Vivian, John.
   The manual of practical homesteading.

   Includes index.
     1. Agriculture—Handbooks, manuals, etc. 2. Organic farming.
3. Agriculture—New England. I. Title.
S501.2.V58   630'.2'02   75-6853
ISBN 0-87857-092-6 (Hardbound)
ISBN 0-87857-154-X (Paperback)

Illustrations by Ginny Mair and Lucille Fatzinger

Contents

In Pursuit of a Better Life 1

About ten years ago Louise and I began to see that there was something very wrong with the typical American super-consumer way of life. Not that we were better able than anyone else to predict the environmental problems, overpopulation, and shortages which are now forcing all of us to change the way we live. Louise and I just got fed up with being consumers—commuting to dull city jobs to keep up payments on a gadget-filled suburban house, flashy cars, and the other nonessentials that symbolize success to most people.

So we stepped quietly off the treadmill, cashed in, and went looking for a better life. It took a while, but we found what we wanted. Today our world is an orchard, gardens, and an old house attached to a livestock-filled barn at the bottom of a long hill in the New England woods. Louise, three-year-old Sam, baby Martha and I are one of a growing number of families who have gone back to the land, though few of us originally came from the country. We are city people who have rejected life in troubled urban America of the late twentieth century and taken up a rural existence that is part life as it was in the eighteenth, and part life as it will have to be in the twenty-first century. Homesteaders, we're called. We grow most of our food the natural, organic way without chemicals; we make much of our clothing, pottery, and tools, do all our own carpentry, plumbing and the like, and supply many of our energy needs with firewood and back muscle.

The life is a good one. We find it much more enjoyable and satisfying than our years in the city, though the work is harder and the hours longer. Admittedly, much of our enjoyment comes from the great contrast between then and now—smog versus clean country air, traffic racket versus the call of a whippoorwill. Plenty of old-time farmers would think us crazy to give up an easy life for work they would gladly leave if they could—or perhaps they just think they'd like to leave. Until recently we were fully prepared to see our homesteading life be only a one-generation experiment; we fully expected our kids to rebel against the comparative quiet of clean air and song birds, reverse the trend, and go back to a city they never knew. They still may, but now we aren't quite so sure. Suddenly we find ourselves uniquely prepared to face up to the severe economic, environmental, and resource depletion problems facing the world. For one thing, on a small farm recycling is a way of life. An organic homesteader works hand in glove with nature, by nature's laws, in the process doing minimal damage to the environment. He is a practicing, practical ecologist, doing what the professors merely talk about. By supplying so many of their own needs and shopping carefully for the few needed purchases, a homesteading family can become almost nonconsumers of the earth's limited or nonrenewable resources. A desirable side benefit of a homesteader's self-sufficiency is his relative economic independence; no matter what happens to the economy no one can fire him or lay him off his own land. And with the garden, livestock, and woodlot, the kids will never be cold or hungry due to the chronic shortages of everything from food to fuel to jobs that have suddenly come to plague us.

THE AIM
OF THE BOOK
The aim of this book is to help others share the satisfactions that we have found by sharing the know-how garnered over the years. In it you won't find a catalog of a dozen ways to do everything as you see in the rash of homesteading books being turned out by city-based journalists these days. We'll cover only the things we do and do successfully. Since we haven't gone all the way back to the preindustrial days there will be nothing on plowing with oxen or splitting shingles with a froe. And we won't get much into clothes making, pottery, renovating and maintaining old homes and farm buildings, or other things we do but which would require books of their own.

The emphasis will be on fostering and using the products of the land in a way that nature intended before technology interfered—living on the land, not off of it, conserving and harvesting its bounty, not exploiting it. We will go thoroughly into organic gardening. The orchard will come in for its share of attention, as will maple sugaring, the bees, and the animals. As for meat, we will cover everything from incubating eggs to butchering the larger animals. And we'll touch on hunting and fishing, gathering wild foods, tanning hides, making fence, heating with wood, and many of the other skills we've picked up in making our lives a partnership with nature and the land.

Before getting into the "how-to-do-it" details, though, I want to answer several questions that people frequently ask—such as how and why we ended up where we are, and where we get the cash to pay the kid's dental bills.

OUR PERSONAL EXPERIENCE

Louise and I are both from the Midwest. We attended eastern universities where we earned bachelor's degrees. After my military service and a few years of working, we went on for Master's degrees, Louise in elementary education, I in business management. After our marriage we began going through the series of transfers, promotions, and job changes that any young business or professional couple endures to advance in income and prestige, pursuing the success that so many Americans consider the prime goal of their existence. We were three or four generations removed from the farm; as confirmed urbanites we were the least likely candidates imaginable for the homesteading life.

However, as our income rose, as we bought sports cars and jetted around the world, we became more and more restless. Part of it was the typical complaint of any city/suburbanite: commuter trains and traffic jams, routine and boring jobs, the needlessly frantic pace of life, and the smog, filth, and crime. But with us the problem went deeper. We were getting no genuine satisfaction out of what we were doing—pleasure yes, but life seemed little more than a search for shallow gratification, comfort, convenience, and status. And though it will be incomprehensible to our elders who suffered through the Great Depression, and to the poor and discriminated-against who are still fighting for conventional success, we found the affluent "good life" too easy. It lacked the challenge of hard, genuinely creative and productive labor that can provide a deep sense of accomplishment. So we stepped out of that world, hopefully to leave our places to those who have yet to taste success American style. Perhaps we can welcome their children or grandchildren to homesteading.

THE BEGINNING

Our departure was in gradual stages. First we sought relief with a series of hobbies, but they soon lost their appeal. Next came various causes, but we found the only one of lasting interest to be the budding effort to save the natural environment. Rachel Carson's *Silent Spring* had just become popular and people were beginning to learn an obscure scientific term, "ecology." Increasingly we began doing physical, so-called menial tasks that our success supposedly let us pay others to do. Louise began making her own clothes again and I tried do-it-yourself projects ranging from fixing leaky faucets to rebuilding an auto carburetor. Increasingly too, we found ourselves listening to the elemental call of the land. First we moved from a city apartment to a town house because it had a miniature patio garden in which we grew tomatoes one summer. Next came a brief fling with the backyard barbeque set and typical suburban living, followed quickly by a move to a lovely little recreational farm with a real barn, located just a ten-minute drive from the commuter train station. With each of these moves the garden grew larger; each signalled a change in attitude, a further rejection of the trappings and demands of conventional success, and a step forward in our unwitting movement to the homestead. More important, in retrospect, each stage enabled us to grow slowly and easily in self-sufficiency and confidence. Before we made the final commitment to a homesteading life we had not only read and dreamed about gardening and animal husbandry and keeping home and equipment going, we had done a good deal of it. Most important, I suspect, we were fully aware that the land does not reward mere good intentions, but demands planning, solid know-how, and long hours of hard work.

THE PIVOTAL MOMENT

Still, the pivotal moment of decision to actually make the change took us by surprise. One spring morning found us in a fancy hotel. I was dressing for an interview for one more new and better job in a new and better city, and Louise was preparing for a polite but ruthless grilling by my new boss's wife. As I stood looking out at the newly-budding trees, tying a good conservative businessman's necktie, I suddenly realized that I hated neckties. I simply detested having the foolish things choking me five days a week and always had, even if I wouldn't admit it. And furthermore, I couldn't get up any genuine enthusiasm for this new job; all I wanted to do was go home and plant peas.

I tried to get in the right frame of mind, tried to adopt the sincere, incisive look and attitude of a hot-shot young business exec. But I kept having this mental picture of myself in old clothes, standing in the garden and grinning like an idiot at a row of just-sprouted pea vines. When I told Louise what was happening to me, her grin grew bigger than my own. It only took us five minutes to decide that we didn't have enough money saved, lacked much of the know-how, and had none of the equipment needed to take up real homesteading, but we were going to do it anyway. Happier than we'd been for years, we threw the necktie in the trash, cancelled the interviews, went home and planted enough peas to feed our whole neighborhood.

Finding a Homestead

For the following six months we kept working at our jobs, saving what money we could and planning for the move to homesteading. By the time we actually picked up stakes we'd found our new home, arranged its financing, found the income sources we would need to supplement our small savings, and generally had the financial aspect of our first two years in the new life fairly well in hand. Still, it was largely by chance that we found an ideal location for a small farm. In the process we stumbled on an unusual formula for buying country property that you might want to try for yourself.

My job had taken us to the Middle Atlantic part of the country, but we wanted to settle in New England where we had spent our school years. We liked the rugged countryside, the lovely rural towns, and the ornery independence of the people. Besides working full-time Louise and I were maintaining a large garden, so our forays into New England were pretty much restricted to rainy weekends. Since it was a long drive North, our calls to realtors were made either late Saturday or early Sunday morning. We always apologized for telephoning at such odd times and then were forthright in telling what we were after—a small hill-farm with a solidly built house and barn, but lacking any electric trash compactors or cosmic-ray ovens or other fancy frills. And we didn't have much money to spend, which meant a low commission for the realtor.

We became convinced quickly that there are two kinds of realtors, those mainly out to make a lot of money and those mainly interested in matching up people and homes. In the latter category you will find good, honest folks who will tell you if a barn has a rotten sill (if they know, which they usually do) or if the seller is asking too high a price. In the first category there must be a few honest people, but there are also a lot of crooks.

We found that there is nothing a realty sharpie reacts to

DEALING WITH REALTORS

with less enthusiasm than someone asking him to spend weekends driving around country roads in a pouring rain looking for a place that won't net him much of a fee. We were turned down by a good many realtors whose names we gleaned from the Yellow Pages and the catalogs of nationwide rural property sales firms. And by the way, it shouldn't have come as the surprise it did, but the most attractive-sounding properties in those catalogs had long since been sold. Remaining good-sounding ones had a feature such as a location next to the town dump that was conveniently missing from the catalog but which the realtor was only too happy to mention on a cold and drizzly Saturday evening.

But then we called a lady member of the Strout organization who not only was happy to talk with us, she even invited us to her home for Sunday morning coffee. After a lengthy discussion of what we wanted she drove us right to it: a big patch of woods with some three acres of well-sodded cleared land on a hillside where water and cold air drainage would be excellent. There was a medium—sized barn in excellent shape and an old house with hand-hewn beams and antique windows that made up somewhat for the fact that the plumbing and wiring also belonged in a museum, the roof leaked, and the front porch threatened to fall in during the next high wind. We took it.

Admittedly we were lucky to find the right place so quickly. But I'm convinced that to find a realtor who combines a genuine interest in your property needs with a thorough knowledge of what's available—which takes a willingness to do plenty of hard leg work—our formula will help. Go looking on weekends, call early or late in the day, and pick the worst weather you can.

Head for the Hills

SELECTING
A REGION

I'm convinced that the best location for an organic homestead is abandoned farm land on hilly, relatively poor land. One reason is the drainage advantage mentioned earlier. Another is cost. Unless it is near a city or resort area, a piece of rugged hillside unsuitable for modern agriculture won't cost a great deal. If abandoned prior to the late 1940s the land will contain no chemical fertilizer or pesticide residues. And taking a look into the future, I suspect that homesteading good arable flat land may be a waste of time, though a good investment. The reason? Sad to say, the world's population is predicted to double in the next twenty-five years. And how will the added millions be fed? The only nation with a significant amount of good farm land not in full production is our own. I agree with those economists who predict that our nation will rapidly become the breadbasket to a world where skyrocketing population growth will push export food prices beyond comprehension. Perhaps eventually rich flat land will once again be worth more as crop land than as parking lots, factories, or sprawling cities like Los Angeles. Land taxes will have to rise too and they could force a flatland homesteader to sell his place to the giant agribusiness firms, though he may be able to retire in luxury from the proceeds.

Our own neighborhood, like much New England hill country, was once heavily populated with small but prosperous family farms. In the early days labor was performed by men and animals that could easily negotiate the slopes and plow around the glacial boulders that crop out everywhere. But after the settling of the rich, deep loam of the Midwest and Plains states, and the post-Civil War introduction of laborsaving machinery, farming the small stone-wall-ringed fields on rocky hillsides became less and less economical. By the 1940s there were only a hardy few clinging to the more hilly land and they were finished off by the introduction of chemical fertilizers and pesticides that made monoculture of huge tracts with gigantic machines first possible, then essential to profitability in the world of modern agriculture. Today most neighboring places are cellarholes where abandoned farmhouses have burned or just fallen in from neglect. Most of the one-time fields, still aching to grow crops, are grown up in timber, the only evidence of the labor of the early settlers being mile upon mile of stone walls running aimlessly through forests of maple and pine.

Sam, Louise, Martha and some rewards of homesteading.

Twenty years ago, says our friend the realtor, places like ours were a glut on the market. An unrestored eighteenth century farmhouse and barn in good structural shape but with the original antique flues and woodwork mostly missing or damaged, and needing modern conveniences such as indoor plumbing, would often sell for less than a thousand dollars. Usually these places would go partially furnished, including old farm equipment and ten to a thousand acres of land, when the last aged member of the farming generation died and city-dwelling heirs sold out as quickly as possible.

Finding such a place in New England these days requires a lot of luck, a fast down payment, and in excess of $30,000—a figure that has much more than doubled in the few years we have been here. Where prices will be in another decade I shudder to think. But ancient New England farms aren't the only possibility. We know of a young couple who spent only $3,500 for a little abandoned wagon factory in an old pasture beside a roaring mountain stream. They have a lot of work to do, but in time will have a fine homestead. And there are unoperated, neglected, or abandoned small farms in odd corners of nearly every state in the union. All it takes is plenty of looking—preferably on a rainy weekend.

Some Straight Talk

Now, how real are the appeals of the homesteading life? Why are we still at it? Is it all it's cracked up to be in proselytizing magazines such as *Organic Gardening* and the *Mother Earth News*? Let's make one thing clear: farm life on any scale is not all roses, not by a long shot. By August every year I get awfully tired of cultivating and Louise usually has some urgent business in town when I finally use up every excuse I can think up and get down to slaughtering pigs. And nobody on the face of this earth can convince me that cleaning out a henhouse is anything but ghastly. However—and this is the essence of the homesteading life—even in the routine or unpleasant aspects we have found satisfactions that have largely vanished from life in modern, urban, industrialized America. So can you. Daily, we experience genuinely productive work. The minutes-old egg that Louise cooks for our son each morning comes from a hen we hatched and brooded and which we house in a pen I built with my own hands. The egg is the product of corn we planted, hoed, harvested, ground and fed the hen ourselves. The shell and any egg Sam leaves will go back to the hen and eventually her droppings back to the land from which we'll grow more corn. And that's another pleasure no city man can ever know—experiencing firsthand the full cycle of life as nature planned it.

THE LURE OF
HOMESTEADING

Perhaps the greatest lure the homesteading experience offers to harassed city and suburb dwellers is its independence from the pressures of urban life and the routine of a nine to five job.

That's a very real satisfaction, particularly to a pair of independent-minded souls such as Louise and me. However, the independence aspect can be overstressed. For one thing, homesteading is hard work, and there is precious little opportunity for a vacation. Many a winter morning, milking goats or feeding chickens, I've longed to be back at a comfortable desk job. And unless you are independently wealthy, which we are not, are willing to live like savages or go on welfare, which we're not either, you need a cash income. To be sure, our cash needs are only a tiny fraction of what they once were, but there are still taxes to pay, kids to educate, and a barn to roof.

Many would-be "back to the landers" dream of living directly off the land, farming or making maple syrup. So did we at first and we haven't given up yet. But with today's land prices and the cost of labor and equipment, you'd need a quarter of a million dollars to start a commercial farm that could compete successfully and support you and your debts. And if you did have that much money, as most any real farmer will tell you, you'd be better off investing it in a bank. You'd earn more.

INCOME

For our income needs, Louise and I quite cheerfully compromise with the desire for independence. At first it was full-time work for us both. Then Louise's teaching job was our only source of income as I put up fence and plowed gardens and began work on several books. But now that the mortgage is paid off we can perform money work on a strictly part-time basis, doing what we both like best. Louise teaches art and pottery a day or two a week and I write, speak, and advise people about gardening, nature, and the outdoors as the opportunity arises. To put a figure on it, we can get along on less than $3,000 cash income a year. That's not to say that a little extra is not greatly appreciated, and is necessary when the truck needs replacement. However, our earnings are usually well under any poverty level you can name. But intentional poverty on a homestead is vastly different from unavoidable poverty in the city. Who's poor if you can serve roast suckling pig for dinner one day, leg of lamb the next, and roast capon the next? With side dishes of organically grown potatoes, squash, peas or beans or corn and a snapping fresh salad. So who cares if the passenger car is a '62 and rusty in places?

NONMONETARY REWARDS

There are other satisfactions. Our kids will grow in a clean, healthy environment and with an intimate knowledge of nature—how the forested hillside changes color as the seasons flow, how goats mate and why corn tassels and where the frogs go

John in some labor of homesteading.

in winter. Always plenty of time to show them how to use a subway token. Perhaps the deepest satisfaction is in living on the land, an integral part of the natural cycle of life as it was intended to function, with the least interference from men and technology. We think of our life as a true partnership with the land, which supplies most of our needs while we try to return to the land in equal measure as we receive.

I don't pretend that we live in a closed system that is totally self-powered and self-sufficient like a balanced aquarium, though we are working on it. Early in our planning I thought I was all for going completely wild and hiking into the mountains with nothing but a backpack and an ax. Louise in her quiet way said I was welcome to try but if I wanted company, hers at least, we'd settle where there were plenty of literate people for her to talk to and kids for our planned pair to grow up with, even if I was content to be a hermit. Also a place where we could enjoy at least the basic amenities of modern life. That meant hot baths and warm rooms for the babies which takes purchased fuel and electricity till we get the capital and energy to install a solar water heater and a wind-powered generator, and to reconvert the central heating from oil to wood.

Still, we do without most of the gadgets and consumer products we once considered essential. As time goes by we naturally find we are getting less and less dependent on machines and mechanical power. Many of the appliances we brought from suburbia have been retired or relegated to part-time jobs. The dishwasher has been disconnected and rigged with shelves, and now serves as an airtight cabinet to let Louise's pottery dry slowly. It's just as fast to hand wash and we use less hot water. Just this year I began scything the orchard grass rather than using the big rotary mower. It too is almost as fast, gives me more and better exercise, and the grass isn't chopped fine but lies in windrows to be raked up for animal feed or piled as a nutrient-rich mulch around the base of the fruit trees. And by heating and cooking with the wood stoves we have just about consigned the propane kitchen range and electric toaster to summer-only duty, while the central heating is needed mainly to keep the children warm during cold winter nights.

The final satisfaction of the homesteading life is being able to pass the good know-how around. Which is what this book is all about—in addition to paying for the new roof. You will find major topics arranged more or less as they come up during the year, starting in January. After a hearty winter breakfast we'll hurry over the dullest part of any line of work, record keeping. But once that's dispensed with, we'll get going on the garden, and spring is just around the corner.

2

Spring and Summer all Year Long

Viewed from the road in late January, our central Massachusetts homestead looks pretty cold and dormant. Snow is yards deep in the fields and surrounding woods, and the buildings are shuttered tight against the bitter nor'east winds. But that's outside. Inside the main house we enjoy spring or summer just about all year long. Even the morning chores offer a trip in memory through the warm days and pleasant labor of the summer past. After a first mug of steaming coffee I'll go down the cellar steps and follow the spring that flows under the floor and into the cold storage room. There in the constant chill stand shelves of multicolored jams, jellies, pickles, and conserves, each jar a reminder of the growing, picking, and preserving that went into it. The glistening bottles of strawberry and dandelion wine, the containers of honey and maple syrup are reminders of hours in the sun and each sack of dry beans or grain, each bin of potatoes, apples, and root vegetables speaks of sweat on the brow, the feel of earth between the fingers, and sore but contented muscles.

After filling a basket with the day's groceries I return to the kitchen where the children and small dog are warming themselves in front of the iron stove, fed with wood cut and split two autumns before. On it Louise is frying thick strips of bacon—home grown, slaughtered, cut, and smoked—as she waits for me to fetch the morning's eggs. I put on the heavy coat, hand stitched from hides tanned in the cellar, and lace up the barn boots—they're bought but the new laces are of rawhide made from a woodchuck skin.

Then it's out past the loaded farm freezers to the good animal odors of the barn. I give the goats and rabbits their morning water and fork down hay that smells sweetly of the sun and summer rains. Then I check the chickens' self-feeder and water fount, collect the half-dozen or so eggs from the back door of the nest boxes, and fill the grain cup to coax the goat out onto the milking platform. After filling the pail with streams of steaming milk and leaving it by the kitchen door to cool, it's back inside for a breakfast featuring minutes-fresh eggs, crisp bacon, and thick slabs of home-baked bread with jam or honey.

So begins the homesteader's day, almost a reliving of the past growing season. And this vicarious enjoyment of the warm months doesn't stop with breakfast. Indeed, we spend much of the winter reviewing the past season's work in the unexciting but "first priority" task of record keeping.

Getting the Records in Order

I realize now that the day we began keeping these records was the day we changed from folks who were just trying out an alternative life style to serious and committed homesteaders. Through the early part of our ten-year transition from city to suburb to ex-urb and finally to self-sufficient all-organic homestead, gardening was primarily recreation. The few precious hours spent working the soil offered a refreshing change from the drudgery of a nine to five city job. It didn't matter a bit how much the garden produced or what the food actually cost. We enjoyed picking flowers and eating what vegetables and fruit we wanted and enjoyed even more giving the surplus to nongardening friends. Gardening was a hobby, and a fine hobby, too.

Not so any longer. To be sure, we still get great pleasure from the gardens, orchard, and animals. It is the genuine satisfaction of providing our most elemental needs through hours of hard, often unpleasant labor. The better the planning, the more efficient will be our use of time, so the more productive will be the land and the more satisfying the labor. Thus the detailed records.

NECESSITY OF RECORDS

It is no small task to collect all our records of last year's food production. In the rush of planting, harvesting, and storing, a lot of information has been scrawled on odd scraps of paper, the backs of seed packets, or the barn walls. The number of hay bales thrown down for the goats was recorded in notches cut in the old hand-hewn beams of the loft, while sale and purchase receipts were "filed" everywhere from the glove compartment of the pickup to a nail in the wall beside the kitchen door. I collect this odd assortment of information, then sit down to transfer it to the permanent ledgers.

OUR RECORDS   Through the year we try to keep up with the following four informal journals.

1. Daily Diary. In the ell—the room connecting house and barn—I keep a large appointment-type calendar with space to write brief daily entries. On it each evening we jot down important weather information, nonroutine tasks performed, and really major events such as the arrival of the bluebirds.

2. Input records. A sheaf of papers, one for each crop or animal, is kept in a folder that travels all over the place with us. There we record all "inputs" made, such as planting information, animal breeding dates, etc. The papers get pretty dog-eared by harvest time.

3. Output Records. This is a writing tablet hanging on a nail above the kitchen sink. Each day we enter details of the harvest—so many ounces of spinach and from which garden plot and row, pounds of milk from the goat, number of eggs, and so on.

4. Garden Plan. This is a drawing to scale of each garden showing which plants went where, planting and harvest dates, and other odd information such as which pieces of land received manure or row compost during the year. It shows the garden reality, vastly different from the plan lovingly designed and redesigned over the previous winter.

Through the late winter I transfer all these jottings to permanent records. Farm bureaus and the like sell elaborate forms for record keeping but mainly I just use plain writing paper. For each goat, for example, I make four vertical columns: the date on the left, followed by pounds of milk given that day, feed consumed, and finally vet or other costs. On the right margin I leave about a quarter-page space for analysis or comments. It is good to know why a particular thing came about and the records give us the information which, with a little thinking, can help us learn from our mistakes or give ourselves a bit of a back pat for an earned good performance. For example, if the goat's record indicated a fall-off in milk production in mid-August, the weather data might indicate a bad heat wave, the feed records might show that we had changed grain supplements, or the harvest records could show that we had slaughtered the last of her buck kids and she was feeling a little lonely.

The final step in record keeping is to tote it all up and figure the comparative value of the several farm operations. I use a hypothetical figure for our labor—the ''wages'' we earn if you will. With the goat as an example, milk production for the year is totaled and multiplied times the price we'd pay for commercial cow's milk and cheese. To this is added similar values for meat and hides of kids slaughtered—the meat valued at retail price of spring lamb, though we think it far tastier. From that I subtract the costs of any feed we had to buy, the market price of feed we grew, plus the depreciation costs on pertinent equipment—the swinging windows in the goat pen, goat-high fence, etc. The resulting figure, ''profit'' (or ''loss'') I guess you'd call it, is divided by the time spent milking and in general care; for the goats, about a quarter-hour a day, which includes cleaning milking gear and an extra half-hour for stall cleaning and maintenance on weekends.

RECORDS ANAYLSIS

The resulting figure, the ''wage'' I'm earning, varies with time and between various tasks. Low point to date has been a good dollar an hour ''loss'' to a top wage of a few dollars an hour. Needless to say, we aren't getting rich. But figuring the cash value of homesteading labor isn't the point of our record keeping. If we were all that interested in money I would still be commuting to a city job. The records and ''wage'' figures are to give us a handle on how we are doing, to tell which activities might stand streamlining if we care to do so. So far the least ''profitable'' homesteading activity has been deer hunting. Each fall I spend the better part of two weeks hauling a recurved bow around our woods looking for deer and only rarely get a clear shot. Considering the cost of all the equipment, I lost about a dollar an hour. Theoretically the most ''profitable'' use of time would be picking watercress from the little brook across the road—over $100 an hour assuming there was enough cress, which there isn't, and we were so inclined, which we aren't. Conventional economic wisdom would say we ought to give up deer hunting altogether, flood the upper pasture, and grow ten tons of watercress each year. I'd say that conventional economic wisdom is a lot of what's wrong with the world. What we value is the health, contentment, and independence we earn, not the money we are missing.

HOMESTEADERS WAGE

A Few Details on One Record

The one record I think it worth going into in some detail is our plant variety record. For each planting of garden or farm crop I record appropriate dates, planting distances, cultivations, and amount and quality of yields. Production is first indicated in standard terms—number of ears of sweet corn, for example. More important to us is yield in what I call "use units," family meal servings or daily livestock feed requirements. We know, for example, that if the weather is good a six-foot row of early spinach will produce three to four pounds (six to eight meals) of a vitamin-rich green vegetable that we like very well. The same land in sweet corn will turn out no more than a half-dozen ears, just one and a half meals the way we eat the stuff. This "use unit" information permits us to decide how much of everything we want to supply our year's needs, then to divide up the land to grow just that much, no more or less. If it sounds as though we eat a rigidly planned diet, be assured that we don't. The planning never works out perfectly, if only because our tastes change all the time. There are always shortages and excesses, but never really major ones. So far our worst mistake was freezing twenty pints of summer squash. Its consistency, thawed, was too mushy for us after enjoying the crispness of fresh, lightly steamed zucchini or yellow crook-neck. The hogs liked it, though.

Records also let us evaluate plant varieties so we can pick those best suited to our ornery New England weather. We'll go into detail on variety selection in later chapters, but one example will show the value of good record keeping. In 1969 a small planting of a normally excellent yellow hybrid sweet corn, Harris Company's **Gold Cup**, turned out with large, tough, and sour kernels. The normal reaction would be to blame the grower. However, records showed that the first good rain in weeks occurred just after the **Gold Cup** went in. A tough, nonhybrid white corn I had planted two weeks earlier apparently didn't germinate till that rain. So despite the planting interval, the two corns must have tasseled at the same time, though I didn't notice it, and cross-pollination ruined the delicate constitution of the hybrid. Without records I would have discarded the seed and never bought the variety again. As it was, the following year we planted double the amount and it produced the usual crop of superb eating. The records also taught me to give a good soaking to any corn seed planted during a dry spell.

Germination Tests

If after a fair test a variety fails to satisfy, the extra seed is discarded and the variety goes on the ''never again'' list. There are a good many on the list and we have given up completely on melons and a good many other plants. No fault of the seed. Blame our short season, shallow, rocky, and acid soil or growing methods. Leftover seed that is kept, though, is subjected to a germination test.

OLD SEED

There are a number of opinions on keeping seed over winter, and I disagree with most of them. First, we have naver had the success some claim for home-raised seed. We have saved some from time to time but have never been satisfied with the produce; like most Americans, I guess we've been spoiled by the success of plant breeders—or the success of those who work to perfect tastier, more healthful varieties for the home garden (their colleagues who are developing square, rock-like tomatoes for mechanical harvesting and perfecting ''green revolution'' corns that threaten the plant world's genetic diversity we don't go along with). We've found that home-bred seed reverts to a rather coarse nature. Perhaps if only one nonhybrid variety of each vegetable were grown so as to avoid cross-pollination, home-grown seed would be fine. But even so, many plants will cross-pollinate with wild cousins, and continual selection would be necessary. For example, garden carrots will happily accept pollination from their wild cousin Queen Anne's Lace, and without careful selection will revert to type in generation or two. Wild carrot is fine as a survival food or an occasional wild food experiment, but the two-inch-long pencil-thin root doesn't compare with **Red-Cored Chantenay**, in my book at least.

It's also debatable whether one should save unused purchased seed from year to year. Many hobby gardeners throw out the extra, which is fine for them and for the seed sellers. Many of these same gardeners try out every new variety that appears in the seed catalogs and that's OK for them too. We did it once too. Now though, we plant to eat. We don't spend money we don't need to, even if we have it, which is seldom. So we save good seed and use it.

Most small, garden-size seed packets have a notation, "Packed for 19--." This does not mean the seed is good only the year indicated, though the seed seller just might like the buyer to think so. Most seed will remain viable for several years and I have read of scientists who have grown flax seed taken from an Egyptian pyramid, and grass seed wind-borne to an arctic glacier that grew after many thousand years in the deep freeze. The homesteader thinks in somewhat shorter terms though, one to five years. Our experience suggests that the larger the seed, the shorter its storage life. I find the tiny seed such as that of members of the cabbage family or carrots are good indefinitely, which is grand since there are thousands of seeds in a packet. Just to be safe, though, we seldom use seed for a major planting after more than three years. The somewhat larger seed such as tomato or radish seems to be good for about three years but we only keep it for two. Corn, peas, and beans are variable, though we don't use even one-year-old seed unless it does very well in the germination test. We have stopped saving the very large seed such as squash, although we seldom have any left over since the dozen or so seeds in a small packet are just right for our year's requirements.

A major problem with saving seed is that you don't know how old it really is when purchased. Seedsmen save seed over winter too, you can be sure of that. On packets near the "Packed for" date is usually a percentage figure showing how many seeds were good in a recent germination test. With reputable companies I've never found these test figures to be overly misleading—though the 89 to 99 percent germination rates you see represent a test under ideal conditions. For years after the one for which the seed is packed and guaranteed, the home gardener must do his own testing for reliability.

We use a pair of shallow metal baking trays. On each go several thicknesses of paper towel. The seed packets are divided into lots—one for the cool-germinating small seeds that will be planted before the last frost in spring, the other containing the tomatoes, beans, and other warm-weather seed. Each packet is numbered and the numbers are written an inch apart in waterproof ink down the long side of the trays. The towel is moistened well and seed from packets whose numbers correspond to the numbers on the towels are spread thinly out across the tray.

Around each tray I wrap several layers of the plastic film drycleaners insist on putting around your clothes. The tray of

To *test viability of saved seed I first draw out a numbered grid on several thicknesses of absorbent paper toweling. Seeds are arranged on this moistened towel—on the "hot" or "cold" tray, depending on their preferred germination temperature. After a week or so we inspect the seed; in this test, all the seed but number 11 [a hot pepper seed] came through fine. And the sprouted seeds make a good addition to the evening salad, so long as you are 100 percent sure they were not "treated" with chemicals.*

small seeds goes into the fifty-degree cool and darkness of our cellar, the other warm-loving one into a cabinet in the baby's room, the only part of the house kept at a fairly constant seventy degrees, even in the winter chill.

Someone checks the trays daily to be sure the seed remains moist. In a couple of weeks every seed that plans to sprout will have done so. Many seeds put out a surprisingly strong, white root that may push the seed itself well out of its row, which is the reason for leaving a good inch between rows. It is quickly apparent which seed is still good. Occasionally germination may be less than complete, but still sufficient that the seed can be used if planted more thickly than usual. If this is the case, a notation is made on the box or packet.

The Seed Catalogs

After the germination tests we know what new seed will be needed and the seed catalogs get a second reading. I'd be the last one to deny the entertainment value of a seed or nursery catalog on a dreary winter evening. We get them all, and I spend many an hour gazing at all those full-color portraits of perfect apples and corn ears with little drops of dew on them. But when it comes to actually ordering seed that is to become next year's food, the catalogs get a hard examination.

SEED SUPPLIERS

We find that the most reliable seed and planting stock comes from the smaller firms which sell primarily to a regional market. Most are family owned and operated, and they don't waste a lot of money on flashy promotional campaigns, ads, or catalogs. Their home garden catalogs concentrate on factual descriptions of seeds or plants with a minimum of color photos and cute copy. They will offer a good variety of each plant or seed and usually make an offer to send a price list for commercial (farmer's) sized bulk lots. Too, the firms we have had success with so far sell either plants or seeds, but not both. This isn't to say that a combined seed and nursery operation can't work, we just haven't run across one yet. I'm sure some of our disappointments have been sheer coincidence. However, we'll continue to stick with tried and true firms catering to our part of the country. J. E. Miller, the mail-order nurseryman at Canandaigua, New York has never failed to satisfy with either plants, service, or guarantee fulfillment. Most nursery catalogs will claim a plant is "hardy" which is meaningless. Miller's says how hardy, giving the degrees of cold various plants have survived. For seeds I stick with our local farm co-op, Agway out of Syracuse, New York, Stokes Seeds, Inc. from Buffalo and St. Catherines, Ontario, and along with the "no-work" gardener Ruth Stout, we patronize Joseph Harris, also from near Buffalo.

The one time we do trade with the highly promotional mail-order houses is when they make their come-on offers of free or low-cost seed or plants every spring. Our March issue of *Organic Gardening & Farming* magazine plus the garden sections of Sunday papers get all ripped up as we tear out the little coupons and send them off, perhaps with a piece of change. The offers, of course, are simply to get the catalogs into your house and the seeds are usually enclosed in the order envelopes of a catalog. (At first I inadvertently discarded several packets of free seed along with what I thought were extra catalogs before I figured that one out.) We find the promotional seed to be adequate, though I suspect it is mostly stuff that didn't sell the year before. Still, we haven't bought French marigolds for three years and we plant them everywhere for bug control.

The final catalog critique is a comparison of format and price from year to year. Not long ago one midwestern firm changed from a simple newsprint catalog with a minimum of color to a real spectacular—a full-color, slick-paper job with more pictures than print. At the same time they reduced the number of varieties offered by half and increased prices by more than half. Their seed was always good, likely still is. But to my thinking they upped prices to mail order customers just to pay for a flashy new catalog. That's their prerogative of course, and I imagine it's helping their business, but not with us.

CHOOSING SEED

But not for a second do I suggest that a serious small farmer or homesteader should buy seed or planting stock on a price-only basis. Cheap seed is sold cheap for a reason. However, price comparison can save limited funds. If two sound firms offer the same amounts of similar strains of a plant at different prices, I pick the lower of the two, and haven't regretted it so far. Speaking of money for seed, the easiest way to save it is to buy from a retail outlet, from the headquarters store of a reputable grower if you are near to one, direct from the nursery store, or from an established feed store selling seed under its own label. That way you save the advertising and shipping costs of a direct mail operation. *But,* don't buy seed from those colorful sales racks that pop up in every dime store and supermarket each spring or seed you find being sold by such well-meaning groups as the Boy Scouts. The seed companies are anonymous—you don't know where they are located, where they get their seed. They may grow it themselves, may get it from any old source, we don't know. What I do know is that we have had universally unsatisfactory experience with it. It's bum seed. Besides, the selection is limited, seed is never specially adapted to a particular part of the country, the varieties offered tend to be out-of-date, the quantity small, and the prices high.

Chemically Treated or Untreated Seed

Dedicated organic farmers and gardeners resist using artificial chemicals of any sort in their horticulture, even the small amounts that are used to coat many kinds of seed. In the past year or so complaints from the organic movement have prompted many seedsmen to offer uncoated seed for the first time in years—though it's just a beginning.

Not all seed treatment is open to criticism. Tomato and other seed can be dipped in hot water to kill spores of several wilts the seed can pick up in growth or processing. Hot water never hurt anything. But several kinds of fungicides are applied to corn, peas, beans, and other large seeds. I've even seen seed packets with the skull and crossbones to indicate poison and others with a notation "CAUTION! Contains Lethal Mercury." Think back a few years—remember the farmer who fed a hog on leftover chemically treated seed? The hog wasn't bothered but the farmer's family was obliterated after eating the pork. Those few who lived are hopeless mental and physical cripples from mercury-caused brain damage. Most treated seed being sold to gardeners is now dyed an intentionally bilious red or green to prevent this sort of thing from happening.

In our experience, the chemicals just aren't necessary on a small homestead. To be fair, a big commercial farmer caught in the clutches of modern chemicalized agriculture needs all the help he can get to maintain his totally unnatural monoculture. In contrast, our gardens of varied plantings are about as close to nature as you can get and though we have some plant disease and bug damage, it seldom gets bad enough to ruin a crop. If it does, it's easy enough to make another planting.

Besides, the seed coatings don't work, at least on our place. Normally our humus-rich organic soil drains so well there is no problem of the water-soaking that causes seed rot. But in spring of 1972 we had rain for two months solid. My early planting of untreated corn seed on the steepest hill garden germinated perfectly. Midseason seed in a lower area rotted, so, more as an experiment than anything else, I tried a packet of chemical-coated magenta-colored midseason corn seed. It rotted too. That year we had our corn early and late, the way nature intended.

Starting Seeds Without Failure

With New England's short growing season we have to start many slow-growing plants indoors in late winter or very early spring. The seed companies purport to simplify the process with some of the most ingenious and expensive seed-starting gimmicks you'll ever want to see. One is a little cube of fiber soaked with chemicals, another is a little wafer of peat wrapped in plastic mesh that swells up when water is poured on. My favorites are the preplanted mini-flats—artificial soil dosed with chemical fertilizers and packed in a little plastic box. Punch a hole in the top, pour in water and grow yourself a crop of plastic tomatoes.

THE GIMMICKS

We stick with more traditional methods. Each fall I mix a garden cart-load of starting medium. One third is sand from a nearby sand bank. Another third is peat moss, purchased in late fall. The last of spring-shipped supplies, it is waterlogged and the package is full of holes, but it is a fraction of the price it would command early in the year. Though the sand and peat provide an excellent loose and porous home for plants' tender young roots, they contain no plant food. Essential nutrients plus even more organic fiber are added in the final third, fully mature compost, the "superdirt" dug from the heap beside the chicken run.

TRADITIONAL STARTING MEDIUM

I sprinkle on a good trowel load of ground dolomitic limestone to counter the acidity of the peat and chicken litter/compost, add rock phosphate to provide the plant nutrient least abundant in our area, and run it through the latest addition to our equipment lineup, the shredder/grinder. Ours is a prototype; the manufacturer asked us to test it, and it is not yet in production. I hope it does get into production, since it has powered wheels and hinged flails like an old-time feed mill. We use the machine mainly for grinding feed. More on that later.

Once it is shredded fine, the starting medium goes into the cart which is run down the cellar stairs where it will stay till spring. Our carts are the kind with two large-diameter spoked wheels. The ads claim the carts are better for garden chores than wheelbarrows or those small-wheeled hopper-shaped things sold in hardware stores, and they are right. We have a small model and one so large it holds a half-dozen hay bales. The Garden Way folks in Charlotte, Vermont make them of plywood with aluminum framing and steel axle and wheels. They are lightweight but rugged. With those big wheels they'll go anywhere with little effort, even down the cellar steps.

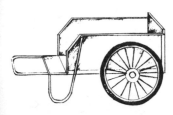

Garden cart

Our total cash outlay for about 200 pounds of starting medium is $1.25. That's for the peat, which I could get free by grubbing it out from the bog across the road. However, we haven't yet been so strapped I couldn't come up with the price of a bale of peat. Besides, the Canadian peat is free of weed seeds; so is the sand, more or less, and the compost after it heats up in the process of natural decay. So our starting medium contains few wild seeds to bring competition to our young garden plants. The same kind of medium, but with chemical plant food, is sold in two-pound bags by seedsmen for about $1.50. That makes what I mix up each fall worth some $150. Louise puts the extra in her window boxes. Who'd believe that each of our petunias and geraniums are growing in dirt that some folks pay seventy-five cents a pound for? Each windowbox contains soil ''worth'' $50, and yet we have some kind relatives who feel sorry for us because they think we're poor.

Avoiding Damping-Off Disease

STERILIZATION   Early in March I fill an old twenty-quart hot-water canner with starting medium and leave it to cook overnight in the kitchen range set at 250°. This is to kill the spores of the damping-off fungus which attacks the stems of small seedlings at ground level. It's as though a tiny rubber band is put around the stem to choke off the plant's circulation. The stem withers at soil level and the little plant dies.

The morning after being heated, the medium is allowed to cool. Then it's put in flats. For this very early planting of small seeds we use flats made from old egg cartons—six by eight-inch cartons made of gray recycled paper fiber that, sad to say, are rapidly being replaced in stores by plastic foam ones. We got a box of several hundred cartons to pack eggs when we have a surplus to **PLANTING FLATS** sell each spring and summer. Most of our customers return the cartons after emptying them, and following several uses they get pretty frazzled. The most ragged become mini-flats.

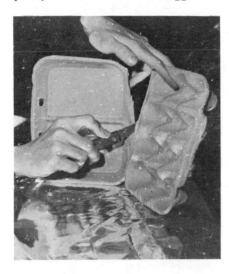

Easy seed-starting flats are made by separating halves of an egg carton...

The halves of each carton are separated, and the two holes in the side of the top half where the "latch" caught are sealed with tape. I pack starting medium into each of the dozen egg cups in the bottom, poke a drainage hole in the bottom of each, then scatter a scant sprinkling of all tiny seeds, such as those produced by lettuce. Two to five seeds of the size produced by tomato or pepper are planted to each cup depending on germination reliability. Finally, I sprinkle on a covering of starting medium—just enough to cover the seed—and pack it down very well. It is best to put only one kind of plant and only one variety in each flat. Combining seed of different germination time, different heat, light, and water requirements is asking for trouble.

The cupped bottoms are set inside their former lids and each remated pair is centered on a seven-by-nine-inch rectangle of aluminum foil. The foil is folded up and cinched over the meeting edges of both halves of the carton, holding the halves together at the same time it creates a perfectly waterproof bottom. I punch a

 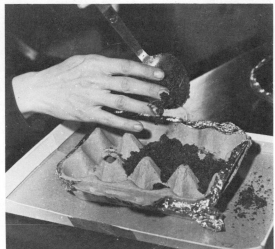

...watering and drainage holes are cut in and a piece of old aluminum foil is sandwiched between the carton top and bottom. In goes the planting mixture of equal parts of loam, peat moss and sand.

hole in one of the egg-separating bumps in the carton bottom where the gray paper fiber protrudes up through the soil and pour in a half-cup of water. The little flats are put into a plastic bag and set away to sprout just as in the germination test: cool-relishing seed down cellar, the tomatoes and other heat-loving plants about the kitchen range.

The flats sit for a week for more till the seed has germinated. One of us looks in on them at least once each day, since a seedling sprouted in the dark can in just a few hours grow a weak and spindly stem in search of light. Once a few sprouts are showing above the soil level so we can be sure all seeds have germinated, the flats are transferred to our miniature hothouse.

FOIL AND PLASTIC　　One note about the foil and plastic bags used in these flats. The time is gone when we accepted a "use once and discard" attitude toward these products, but they do have a place on the homestead. You can't escape plastic bags; everything sold comes in them, clothes, food, you name it. We save the bags, wash and reuse them several times. Indeed, any time we have to buy supermarket produce or bread, the decision is often made on the quality of the bag. We don't have much hope for the quality of its contents. We wash and continually reuse aluminum foil too. Been on the same roll for five years now. Each sheet is washed after use, folded and stored away till needed again for cooking, sealing the top of a can, or use in making mini-flats. When it is completely worn out, each piece is crumpled in a ball and put into a box for eventual recycling.

A Greenhouse in a Window

In New England during late winter and early spring the sun is so weak and shines so few hours a day it can't nurture plants properly. To have lettuce in May, spinach in mid-June, and tomatoes by July 4, the plants must be provided an artificial version of July weather during the late winter.

I have seen ads for miniature hothouses with electric trays to warm soil, plastic domes to keep in moisture, and movable fluorescent lights for about fifteen dollars per square foot of growing space. Our set-up cost about one-third of that, which still isn't cheap, most of the expense going for fluorescent light fixtures—eight to ten dollars when we got ours.

The location is a window on the south wall of the kitchen, where four shelves, 2½ feet wide by 10 inches deep, are supported by movable brackets. Standard 24-inch fluorescent light fixtures are fixed to the bottom of the top three shelves and to a thin board over the topmost shelf. We have used the special tinted ''grow light'' bulbs, cool white, and the regular bulbs with a yellowish tinge, and can't see any difference.

Shelves are adjusted so the bottom of the fluorescent bulbs will be an inch or two from the growing tips of the seedlings. At first the shelves are equal distances apart and all jammed into the top of the window. But as the plants grow, at different rates, shelves are gradually moved down, and the flats moved around so plants of approximately the same height will be on the same shelves and maintain the proper distance from the light.

I have never noticed any harm to seedlings growing too close to the lamps; they often touch the glass. However, too great a distance will cause spindly growth, making for overly long and weak stems. We leave the lights on long enough each day to provide seedlings at least ten hours of total light. Of course, artificial light lacks the strength and variety of radiation provided by natural sunlight and there is no doubt that the little plants prefer the real thing. Any time the sun shines, even weakly in February, the seedlings bend toward it hungrily.

A half-cup of water is poured into each mini-flat whenever the soil on top begins to look dry. Water is usually needed twice a week if the seal between top and bottom of the egg cartons is good. Drawing water up from below encourages strong root growth,

And here are the flats with started tomato, eggplant, pepper and broccoli seedlings on the adjustable shelves of the "window greenhouse."

which is a seedling's main job the first few weeks. But early in March we notice the little nubbins of the first pair of true leaves appearing between the cotyledonous leaves of an especially precocious tomato or pepper seedling. Then it really is spring in the kitchen, even if the window lights reflect off a solid wall of blizzard-driven snow.

Sweets of Early Spring 3

One morning in early March, often before the seedlings are high enough that we need to make more room between the shelves in our window greenhouse, the sun will rise into a cloudless sky. By 9:00 a.m. the rime frost will vanish from the beams and inside walls of the barn, and by ten a steady dripping will commence as the roof-top snow begins to melt. By noon the temperature will be a balmy forty or fifty degrees and a townsman driving past might be surprised to see us in our shirt sleeves, snowshoeing up the hill to check the beehives and maple trees. Once the bees take their spring cleansing flights and the maples begin to flow, spring flowers are just around the corner. And so are nature's own sweeteners, maple syrup and honey.

After their grand taste, perhaps the biggest attraction of maple products and honey is the cost—free for the taking once you've obtained needed equipment. And compared to refined sugar's empty calories, they are positively good for you. Maple syrup contains a concentration of all the sugars and minerals that occur in the sap which is literally the "life blood" of the maple tree. And honey beats that. It contains small but significant amounts of every vitamin needed in bee (and human) nutrition. Vitamin C comes from the pollen in the honey. Honey's mineral content includes, in order of strength, potassium, chlorine, sulfur, calcium, phosphorus, magnesium, silica, iron, manganese, and copper. There is so much potassium that bacteria can't live in honey—it is actually an antibiotic agent, though by no means a medicine. And there's more. The several sugars in honey are more digestible than ordinary sugar and it is recommended for babies. Practitioners of folk medicine have known for years that infants that cannot take cow's milk and sugar formulas will thrive on goat's milk and honey. And to cap it all off, honey with comb and all is a proven symptom suppressor for hay fever and such sneezing allergies. I should know; for years I suffered with maple pollen each spring and who knows what each fall. Chewing honeycomb doesn't cure it completely but it is effective, and a whole lot tastier than any pill I've ever had prescribed.

Good as honey is on toast or maple syrup on hot cakes, though, I won't pretend we use them in every recipe needing a sweetener. We tried and quickly got pretty sick of their distinctive flavors. Even the single tablespoon of maple sugar used to start yeast in white bread gave us off-flavored bread. Maple sugar also proved unsatisfactory in coffee. After several weeks of using it and getting used to the flavor, it began tasting bitter. I can't explain why. After switching to honey in coffee for a while the maple stopped tasting bitter. But then we got tired of honey-tasting coffee too. One or the other is used in most desserts and plenty of things that are supposed to taste of honey or maple, but we still keep a bin of semirefined cane sugar around.

Sugaring Off

In our area maple sap will run on and off from mid-February to early April so long as day temperatures are well above freezing and the thermometer falls to twenty degrees or lower at night. There are many other weather considerations of importance to our neighbors who sugar-off on a large scale to earn a living. They worry about the amount of sun, direction of the wind, depth of snow around the trees' roots, and so on. They also get to work early in the season since the best sap is suposed to flow in February. The later in the year, the darker and more flinty tasting the syrup. *The Maple Sugar Book* by the original twentieth century homesteaders, Scott and Helen Nearing, contains as much detailed maple information as anyone might need to set up a paying sugarbush. But since Louise and I collect sap only for our own needs, we don't pay much attention to the finer points. When the trees run we collect the sap; when the weather is fine we boil down. And then we never use all the syrup or sugar we make. However, obtaining sweeteners is just one reason for tapping the trees. Mainly, sugaring off is the first good, hard outdoors work of the year. It is better than any spring tonic to get out into the crisp air, tramp from tree to tree with brimming sap containers, boil down in clouds of steam over a rich-smelling wood fire, and then sleep the night with a welcome ache in your winter-softened back muscles.

THE RIGHT TREE

The first thing needed to get maple sugar is the proper tree, the sugar, hard, or rock maple—*Acer saccharum* in the botany books. It takes a little experience to tell a leafless sugar maple in early spring by trunk and branch shape and the gray flaky bark. At first I had to pick the trees the fall before by looking at the leaves. They are five-lobed rather than three-lobed like the only other comparatively common relative, the black maple. The leaves are less pointed than those of the relatively scarce silver or soft maple. The trees grow in uplands, not swamps as do the red maple, and they are big trees, not runty like the mountain and scrub maples. I have read of folks living outside the sugar maple

Red maple

Black maple

Sugar maple

belt who tap other species of *Acer* with some success. Some say a fine syrup can be made from sap of the deep South's sweet gum tree. Anyone living south of Boston ought to look through the wild food books for sugar-making ideas. Just don't tap a pine tree.

YIELD

Spile

Quantity and sugar content of maple sap varies from tree to tree and season to season. We figure it takes about fifty gallons of sap to make a gallon of syrup or eight pounds of maple sugar. If you work at it, a good tree will produce up to twenty gallons of sap per tap. We tend to be a bit casual about gathering the sap and are lucky to get in half that amount. For every quart of syrup or two pounds of sugar we want, I plan on tapping one tree. Last year we still had sugar left from prior seasons, so I only put in a couple of dozen taps. Still, we put up twenty quarts of syrup and gave away fifteen. As I say, there are more reasons than one for sugaring off.

Spile Whittling

After the right tree is found, the next thing needed is a supply of spiles (pronounced with a hard ''i'' as in ''while''). Spile is the proper name for the spout you put in the tree to get the sap out. Indians used to cut diagonal gashes in the bark much as they tap rubber trees in the tropics, but that isn't good for the tree. It's asking for disease problems. Spiles are better. They sell galvanized steel ones in a few New England hardware or feed stores, and I have tried simple lengths of pipe with a notch filed in the end to hold a bucket. The best ones, though, are hand whittled of wood.

WOOD SPILES

First, wood spile doesn't cost anything. It fits tighter so the sap containers won't pull out and spill and practically no sap is lost from leaks around the outside. When the run is over you don't have to pry the wood spile out and next spring what is left of it tells where you bored in the year before so you won't tap near there and hit scar wood which won't drip a drop. Besides, I suspect that maples take kindlier to friendly wood being punched through their bark, though I can't prove that.

Spiles can be made from any wood you are willing to drill out, but easiest are small trees or bushes with hard sap wood and a soft, pithy core. I use sumac, the bothersome weed tree with compound leaves and red fruiting spikes. The new wood, which is fuzzy on the outside, is too soft, so in the fall before heavy snow I cut down several year-old trees about a half-inch through and use the trunks.

MAKING SPILES

There is no better way to spend a cold winter evening than to sit in front of a roaring grate fire whittling spiles. Tools needed are a good pen knife and a ream made of a foot length of clothes hanger wire with a loop handle on one end and the other end hammered out flat and sharp on the anvil. I cut the wood into five-inch lengths by rolling it under the knife blade till I'm through the sap wood. It is soft enough that there is no need for a saw, and cutting with a limb lopper only splits it.

Next I ream out the pith. (Supposedly they did this with a hot poker in the old days. That might be fine for five hundred spiles, but for our needs it would take twice as long and the sumac smoke is a bit tangy for me.) For the end that goes into the tree,I whittle two inches at one end down to a taper, the tip as small as possible and the butt just a little bigger than the auger used to tap the trees. The spout end is shaved to a flat point, and on the under side of the tip I carve out a little curved lip so the drops will fall free and not climb the spile and drain down the bark.

To hang the sap container, I cut a good notch on top, halfway up the spile. The front of the notch should be cut straight down into the wood, then it should slant back toward the butt. That is so the container will slide on easily and then stay put.

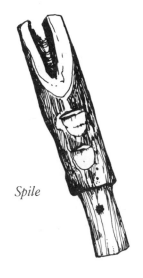

Spile

Besides the spiles, you need a drill of some sort. I use a carpenter's brace and a ¾-inch auger bit. A hand ax is helpful to chip away bark, and of course you need containers to catch the sap. The Indians hollowed out logs, the first settlers used wooden buckets, and their sons and daughters used metal ones. Big commercial operations today use complicated systems of pipes, tubes, and large collecting vats. Louise and I started off using coffee cans with coathanger handles, but they rust and are so small they have to be emptied several times a day during a good run. Now we use the one-gallon, returnable, white plastic milk jugs used by several dairy chains, until the government decided they weren't sanitary enough. The plastic is made to be cleaned and reused many times so it's tough and won't contaminate or off-flavor the sap. The jugs have short plastic handles permanently attached, and if the spile notch is cut in right, they can be hung with the spout end of the spile wedged into the jug opening so little rain can get in and the jug won't blow off in a wind. The only disadvantage of the milk jug comes during a hard freeze. Ice can't be removed as from a bucket. But it's no problem to bring the jugs inside to thaw.

OTHER EQUIPMENT

Before tapping the trees, we wash the jugs out with a mild detergent and several rinses of very hot water. Any spiles saved from prior years are cooked in boiling water for several minutes to

THE TAP

kill the black, spotty mold that will form in any sap hole once the weather warms in spring. To tap in, I find a spot on the sunny, south side of each tree under the largest limb on that side. Morning sun will warm a tree and start the sap run earlier, and for some reason, flow is greatest under the largest limbs. The old wisdom says that the sap flows from the roots up to the tree top, the largest quantity and strongest force going to the biggest limbs. I cut a maple one spring and both the stump and cut trunk ran sap, commencing when the weather warmed in the morning and stopping when it cooled off at night. As I recall, this went on for several days, and when I finally cut the limbs both ends of them ran sap too. So whether sap runs up, down or sideways doesn't really matter to us. When it runs we tap it.

I select a spot at least a foot away from any old tap and chip off the large bark scales so the milk jug will lie flat against the trunk. Then I drill in at a very slight upward angle for about two inches. The hole is cleaned of shavings and will immediately begin to leak milky sap. I let the tap clean itself till the sap runs clear, then put in the spile, snugging well with the flat of the ax but not so hard the spile splits. Then the jug is hung, and shortly our ears are treated to the first musical drip, drip that's the song of a New England sugarbush in spring.

Boiling Down

A good tap will have to be emptied at least once, often twice a day during a strong run. We accumulate sap in an assortment of containers till we have the twenty-five or more gallons needed for a day of boiling down. Sap will ferment in less than twenty-four hours during warm weather, so on balmy days we either pack snow around the collecting vats or put them in the cellar.

Large-scale sugaring operations use wood or oil-fired evaporators, which are relatively shallow pans with a series of baffles and gates attached to the bottom and sides. The fire and flow are regulated so that fresh sap poured in at the top has boiled down to syrup by the time it reaches the bottom. A neighbor two miles to the northwest has such an operation and sells several thousand gallons of syrup each year. At the other extreme are some folks that boil down a few pints of syrup in their electric skillet. We have a compromise.

In the center of the back yard is a big fieldstone fireplace OUR
and cooking pit. The fireplace grate is just right for keeping a EVAPORATOR
sugaring fire going under an old one-by-three-foot black iron sink.
The sink is supported on an arrangement of loose firebrick from
torn-down chimneys. After being washed and rinsed well, the sink
is arranged so the solid end is over the fire pit and the end with the
drain juts out almost to the edge of the fireplace. Under the drain I
put a length of aluminum roof flashing bent into a gutter, and in

*Hammering in a sumac-branch spile. If you look real carefully you
can see the remains of last year's spile—my left thumb is pointing
at it.*

the drain hole I jam hard a two-foot length of maple branch,
debarked and whittled to make a snug fit with the drain opening
and covered with a square of old towel.

We fill the sink half-full of sap and build up the fire of
maple or hickory cut the fall before. As the fire matures and a good

bed of coals develops, we will add all the dead wood fallen in the yard and gardens during the winter.

Metal buckets full of sap are placed around the perimeter of the fire to warm. As sap in the sink boils down, the preheated sap is added, poured in through a wire mesh strainer to remove the odd bits of wood and bark, and later in spring, the bugs that are attracted to the sap. The modern way to boil off is to complete each batch rather than to continually add sap and boil the same syrup for hours. The modern product is lighter colored and has a more refined flavor, I suppose. But we like it the old way, flavored with a

Louise gathers in the sap, while I supervise the boiling down in the old black iron sink.

lot of fly ash and scorch and the occasional chunk of bark that falls in and floats around for a while. The syrup is dark and a little smoky tasting, reminiscent of the crackling fire and billows of steam and the happy labor of making it.

We keep adding sap till the supply is gone or till night falls. Then the fire is pulled out and the hot ashes dumped in the snow up at the garden where they will add needed potash for the

spring planting. The syrup is left in the sink till boiling stops so the sink can cool gradually. It's a good hundred years old and, especially on a cold evening, might crack if let cool too fast. When it cools I put a five-gallon bucket under the drain, pull the plug, and carry the steaming bucket into the kitchen.

WORKING
IN THE HOUSE

The sap is strained twice through several thicknesses of cheesecloth and put in a big kettle on the kitchen range. One of the modern frills we have kept is a powerful kitchen exhaust fan that is built into the old-fashioned wood stove-type top that I attached above the gas range. The fan pulls out the steam as we finish the sap inside, which is a good thing. The steam contains a small amount of sugar, and as people who try doing all their boiling in the kitchen find out, it can coat walls, ceilings, curtains, and the cook with a thin sticky layer of goo. In fact, even though we do most of the boiling outside we find a day's exposure to the sugary steam sours our taste. A lot of dill pickles and homemade kraut get eaten during sugaring off time.

MAKING SYRUP

To get syrup of the proper consistency, we put the candy thermometer in a pan of rapidly boiling water and note the temperature, which will vary somewhere around 211° to 213° depending on weather. The sap is then boiled till it reaches a temperature exactly seven degrees above the boiling point of water. If it was relatively thin when brought in the sap will stay at or near water boiling temperature for a long time. It will rise the first few degrees very slowly but the last few can come quickly. Without frequent checking the batch can be burnt and lost.

When finished, the syrup is strained through several layers of cheesecloth laid in a large-size kitchen strainer, and then bottled. We use the most attractive ''no-deposit-no-return'' beverage bottles we can find, though the contents are poured down the sink as often as drunk. (What better way to reuse glass?) The bottles, needless to say, have been cleaned well and sterilized in boiling water. The bottles are corked immediately after filling with the same hand corker we use in making our wines. When the syrup has cooled, Louise melts a can of paraffin and dips the neck of each bottle in wax. (If you do it when the syrup is hot the wax will drip down over the bottle.) Finally, the bottled syrup is stored in the cool of the cellar to await a pancake or waffle breakfast or shipment along with an assortment of other homemade foods as gifts to friends and relatives.

Making Sugar

To make sugar we cook the syrup another seven degrees to get a moist candy-type sugar, or one additional degree for a dry granulated result. Removed from the heat and poured in a shallow pan, it is let cool to near room temperature, then stirred with a wooden spoon. Shortly it will begin to take on air and as it crystallizes will change from a dark brown heavy syrup to the lighter shade of sugar. You can pack it in molds while still warm if you want candy. We don't, so we let it harden in the pan. It hardens in lumps and tends to pack up, which is fine with us. For a more granulated, refined-appearing sugar we cook it the additional degree of temperature and whip as mentioned above, let it dry well, then grind it up in the blender. It will then store without caking. (In the old days they dumped the moist-cooked sugar into a wooden barrel, dug a pit in the center, and scooped out the syrup that collected there. If anyone wanted hard sugar they would chip off a hunk from the outside.)

First Chores With the Bees

During the sugaring off season, our only task with the bees is to make sure the top section of the hive is exposed to the sun. If need be, snow is kicked away and some of the insulating hay packed around it the fall before is untied and pulled away. Each warm day the bees should emerge from the winter exit at the top to take the brief flights in which they empty their digestive systems of excreta accumulated since the last flight period of sunny, hive-warming weather, which was probably the inevitable January thaw.

Once the bees are flying every day, we have to begin working with the colony. The hive must be opened and inspected periodically and that's when a novice will appreciate any time he's spent with an experienced beekeeper, if only to learn that bees don't sting if handled properly. Some years back I was lucky enough to see my great-uncle handle bees that he kept mainly to pollinate his apple trees. His hives were sections of hollow logs about four feet long, split and retied together. They are called

THE OLD WAYS

Skep

gums, since they originated in the South where the early pioneers used sections of sweet gum, a tree that usually hollows with age. My uncle used no veils or gloves, but after shooing the kids away, he would build a smoky fire at one end of the gum. Alternately fanning smoke over the gum and working with a mallet and froe—a shingle splitter—he split the gum in half and opened the comb inside with one cut of a long brush knife. He stepped back quickly as the bees boiled out, but kept smoke flowing over the gum for several minutes. The secret, he said, was to ''smoke 'em good'' because the smoke makes the bees gorge themselves on honey for some reason. Full of honey, they can't bend double which they have to do to get their stinger working.

When he figured the bees were smoked well enough, the old fellow would cut out fist-sized chunks of honeycomb, enough to fill a milking pail or two. Then after tying the gum back together he would let us boys dive in. Now that is pure joy for a nine-year-old—the sweetness of honey to savor, wax to chew for the rest of the afternoon, and honey all over your hands and down your chin and in your hair so the whole gang had to run down and jump into the creek without bothering to go back to the house for swimming suits.

A few mountain folks keep bees in gums and a few may still use skeps, the hives developed in Europe where the honeybee is native. (Our honeybees, wild or tame, are all descended from early immigrants that came to America with the first settlers.) A skep is made of grass tied together in a thick rope. The rope is coiled up into a half-round dome. I suppose the hive is opened by uncoiling the rope, cutting it away from the comb as you go, but I don't know for sure.

Both gums and skeps are pretty wasteful. Left to their own devices bees build comb in a wavy pattern, intermingling honeycomb with pollen-storage comb and the brood comb used to raise young. Cutting honeycomb away from the other parts wastes

FIRST DECISIONS honey and kills a lot of young bees. Much as we admire many old-time ways of doing things, we have given in to progress and use modern beehives with movable frames for the bees to build comb on. Someone with the time, patience, and precision woodworking tools to work up the 100 or so pieces that go into a hive is welcome to the job. We purchased ours, and they aren't cheap. A hive for only one colony, managed so it will survive New England winters, costs fifty dollars or more. You have to like honey a great deal to spend that kind of money, and actually we didn't till we tasted some ''home-grown'' from a neighbor's bees. It is murky, not filtered clear like commercial honey, and each batch has a different color and flavor. The years we plant buckwheat we get a honey that is dark purple almost, with a nutty flavor. Spring honey from maple blooms has a butter-maple flavor, apple blossom honey has a wonderful aroma, and the honeys from wild flowers such as clover, milkweed, and goldenrod are all different and delectable.

Hive Location Is Important

Before sending in the first equipment order, a homesteader should be sure he lives in an area that produces nectar and pollen in sufficient quantity to support colonies and produce surplus honey for his own use. ''Honey for Sale'' signs in front of farm houses in the area are a good sign. So is the presence of a lot of wild bees—our fruit trees were literally buzzing with bee activity before we set up the apiary. However, any homesteader in a relatively arid locale, the extreme North, or areas where a lot of crop spraying goes on, might think twice before taking up beekeeping. Pesticides kill millions of bees annually, and who knows how much gets into honey?

Next, in the North at least, the homestead must have locations for hives which offer, in order of importance, 1. shelter from winter winds; 2. shade from mid-summer sun; 3. easy access to fresh water and nectar-producing plants. The location should also be as far as possible from yards where children play or animals graze. Bees do fly in a ''beeline'' from flower to hive. Anyone standing in a well-used flight path can have an experience like mine the time a bee in flight collided with the back of my neck (with a suprisingly hard thump), and fell down my collar to buzz around furiously before inserting its stinger into the small of my back. Louise said she thought I had come down with a sudden case of St. Vitus' Dance.

LOCATION CRITERIA

We are fortunate to have nearly ideal hive locations under the maples growing along the stone wall at the north end of the orchard and gardens. A thick stand of pines grows beyond the maples, and the stone walls and evergreens break the prevailing northeast winter winds. The maples shade the hives in summer, though the leafless boughs in winter permit the sun to get through and warm the hives. Seasonal brooks run down each side of the orchard so water is available most of the time. During any unusually dry fall the brooks will dry up, so we provide water in a section of wooden storm gutter with the ends sealed with aluminum roof flashing. Several rough stones are put in the trough to give the bees a landing place and good foothold; without them many bees would fall in and drown. The hive openings face south toward the orchard and gardens, and with as many as a quarter million flying pollinators at work we have very few infertile blooms on any food-producing plant.

Equipment

HIVE DESIGN

The modern hive is based on the principle of beespace. A bee demands a precise amount of room to move around in, 5/16 of an inch, give or take a sixteenth. Any less and the bee is excluded, gnaws out the necessary room, or fills the space with propolis, a glue made from tree gum. Any more space and the bees will fill it with comb. A hive is designed to get the maximum amount of comb into the minimum amount of space. It is a wooden box about twenty by sixteen inches with rails along two facing sides. From the rails are suspended (usually ten) rectangular wooden frames holding thin sheets of beeswax comb foundation. The bees build comb on each side of the frame and have perfect beespace between combs and at the top and bottom of the hive. So long as beespace is maintained, the bees will not glue the frames together too badly and the beekeeper can remove them to take honey or inspect the hive.

USED HIVES

We had a chance to buy used hives at a fraction of the price for new but decided against it. If someone has stopped keeping bees there is a good reason. If the reason is bee disease, a beginner may be licked before he gets started. The worst bee ailments are three diseases where the young larvae die and sort of melt in their cells. Most serious are American and European foulbrood, while the third, sacbrood, usually does only moderate damage and goes away by itself. Fortunately we have never experienced any of the three—I say fortunately because the only sure-fire way to eliminate foulbrood is to destroy the colony and burn it, comb and all. The frames and hive bodies must be carefully sterilized. In fact some states legally require such treatment, require annual inspection, and offer professional assistance. To find out state requirements, I'd advise any new beekeeper to send for "Beekeeping in the United States", Agriculture Handbook No. 335, which is available for $1.50 from the Superintendent of Documents, U.S. Government Printing Office, Washington D.C. 20402. Besides the data on state regulations it also contains almost 150 pages of solid information on all phases of beekeeping.

The major disease of adult bees is nosema, which affects overwintered colonies in the North. After especially severe winters we prevent it by feeding the bees for a week on a spring tonic of sugar water containing Fumidil B, a medicine developed by the federal government specifically for this bee disease. Commercial apiaries, which keep colonies in unnaturally dense concentrations, incur many other diseases and have an arsenal of drugs and poisons to combat them. However, you can be sure none of these problems will exist in new hive equipment or with your packaged bees. And they will seldom crop up in a small homestead-sized apiary.

SUPPLIERS

Three good bee equipment makers are Dadant and Sons, Inc. of Hamilton, Illinois 62341, the Walter Kelley Company of Clarkson, Kentucky 42726, and the A.I. Root Company, Medina, Ohio 44256. All are happy to send free catalogs, all make and sell comparable equipment, though Root is most expensive by a small amount, Dadant next. We buy Root equipment because they have a retail outlet nearby and the savings in mailing costs make up for higher priced goods. I can't speak for the other manufacturers, but Root equipment is beautifully machined. If you like working with wood, you'll appreciate how well the mortises join. And everything received from them, whether wood, metal, or paper, smells like honey.

NECESSARY EQUIPMENT

All manufacturers and many mail order firms in farm and garden-related businesses sell a starter kit, "all you need to begin with bees," for thirty to forty dollars. I'd say about half the equipment is a waste of money. You get a standard hive (20 by 16 by 9 1/2 inches deep) plus frames and foundation, which are essential. So is the bee feeder and bees, but they cost another ten to twelve dollars, mailing costs from the southern apiary included. But the special bee gloves, bee veil, smoke blower, hive tool, and instruction book aren't necessary.

The first time I handled the bees I was as scared as any novice and wore a pair of gauntleted horsehide work gloves and my

The neighbors help with honey-gathering. Here Kip Corey removing a honey-filled super.

mosquito net used mainly for spring fishing. With experience, though, comes confidence, and for a while I didn't use any protection at all except when tearing down a hive completely, which takes a good half-hour and disturbs every bee in the colony.

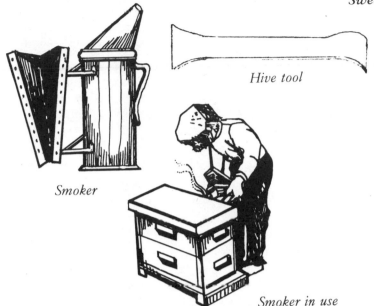

Hive tool

Smoker

Smoker in use

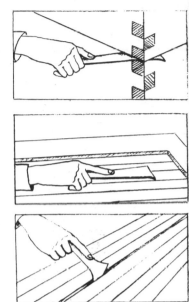

Hive tool in use

The occasional sting I received caused no problems. Then one cool afternoon I had to show a young visitor how accomplished a bee keeper I was. Bees get grumpy when it is too cool for them to work, and when I removed the hive top a squadron boiled out. With a good half-dozen stings around head and neck I swelled up like a red balloon, had to have medical treatment, and am still undergoing a series of shots to desensitize me to bug stings. In retrospect, I should have had the simple allergy test that reveals this common allergy, which can be fatal, before taking on the bees. I'd certainly advise any potential beekeeper to do so.

The hive tool is a miniature wrecking bar but I've found my stockman's claspknife fine for all bee chores. The smokeblower is a small bellows attached to a fire chamber. You are supposed to keep a smoldering fire of oily rags or corncobs going and blow little wisps of smoke into the hive opening to upset the guard bees and keep the others down in the hive while you work on it. I did buy a smoker, but found it did more harm than good—made me cough and sneeze, which bothered the bees. I'm sure I'd feel differently if we had a big operation and had to work the bees every day, whether it was good bee weather or not.

All manufacturers sell "how to" beginners' books which I find confusing—*Homo sapiens* has been studying *Apis mellifera* for centuries and there is a lot of information in every bee book that is of interest only to bee scholars. I'd say, get the government book mentioned earlier, and when actually ordering equipment, get *The ABC and XYZ of Bee Culture* by the A.I. Root people. It costs seven or eight dollars and contains everything you need to know and a lot you don't in its 700-plus pages. But once you get a colony going you'll find it essential in a practical sense and just plain fascinating reading. Would you believe that the length of bee tongues varies with latitude of their range? Or that the average bee flies at only fifteen miles per hour?

GETTING STARTED

In the South, the Northwest, and other balmy climates, bees can be started nearly any time of year since there is a more or less continual honey flow. Florida beekeepers are packing orange blossom honey, those in Oregon are collecting apple blossom honey while we are still buried under ice and snow. In our cold climate we had to start the first colony as early as possible in spring so as to have a strong population of young forager bees ready when the first fruit blossoms opened in mid-May. Since worker bees take three weeks to develop from egg to bug and spend the first three weeks of their life working inside the hive, that meant we had to have everything ready to go by the last week of March.

BREEDS

Before mailing off our order we had a final decision to make: which strain of bees to adopt. Just as plant breeders have developed **Butter & Sugar** sweet corn from a native that was little more than an overgrown head of grass seed, bee breeders have developed strains that are more productive and gentler than the wild stock. These hybrids go under various trade names but almost all sold in the United States are one of two varieties, Italians or Caucasians. The Italian bees are most popular and the strain we decided on. They are fairly easygoing but productive; they are hardy and not inclined to swarm as much as the Caucasians. This latter strain also tends to gum up the hive with propolis, particularly in the South. They are extremely gentle, though, and a good choice for suburbanites or homesteaders with close neighbors.

Packages of workers and drones come in two-, three-, and five-pound sizes with or without a mated queen. We changed our initial money-saving two-pound order when we read that the larger sizes are essential to get a strong colony started for first-year honey production. The two-pound packages and all queenless lots are for strengthening winter-weakened colonies. They also sell lone queens for replacing a dead or inferior queen in an established colony. Before reading up on bees I thought we could really get started cheap by buying only a queen. This doesn't work, as the

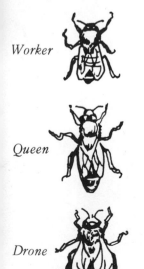

Worker

Queen

Drone

queen is helpless without thousands of workers to build comb, care
for the young, defend the colony, and gather food.

The Hives Arrive

The initial excitement we felt when the long, flat boxes of
knocked-down hive parts arrived paled a bit when they proved to
contain hundreds of pieces of wood and nails in six sizes. Sam was
the baby and he thought it was a new set of blocks, and with his
"help," assembly took hours. The job isn't particularly difficult
and directions are simple, explicit, and must be followed to the
letter. I got a bit too casual and put several frames together the
wrong way around.

On page 48 is a sketch of the hive we have found best for
our locale. The parts are explained on the facing page. The hive
should be painted a light color to reflect summer sun. The
manufacturers sell a special hive paint at a premium, but any paint
will do so long as it is waterproof when dry, and doesn't contain
lead or other poisonous metals. We used leftover white latex house
paint.

The only other essential equipment are bee escapes, queen
excluders, and the feeder which is explained later. The excluder is
a grill through which workers can move freely, but which keeps
out the queen and drones with their larger abdomens. Placed
between brood frames and the super, it keeps the queen from
laying eggs in the honey that will go on your toast. Excluders
come in several styles. Bees prefer walking on wood, so the best
(and most expensive) are excluders with sets of parallel wires set
into wooden slats. We have two excluders. Once in a while the
bees will refuse to enter a super, so I put an excluder on top and
bottom, then place the super between the hive bodies. The
workers have to go through the super to get to brood in the upper
hive body, and quickly get used to the super.

Bee escapes are oval metal stampings with little springs
over an opening that act as one-way doors. Placed into the hole in
the inner cover which is set under a full super, they will clear the
super of bees in a day's time. Optional and unnecessary for a small
honey operation is a whole battery of expensive equipment for
getting "extracted" honey. This is what you buy in bottles in
stores. You need a five or ten dollar electric uncapping iron to get

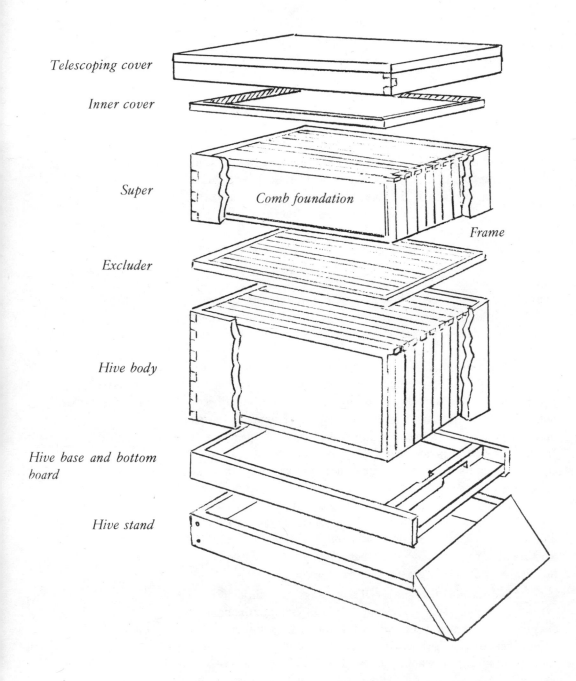

Telescoping cover

Inner cover

Super

Comb foundation

Frame

Excluder

Hive body

Hive base and bottom board

Hive stand

Hive Stand: *A rectangular frame made from nearly rot-proof cypress wood. Keeps hive off the ground, away from damp. Front piece is about five inches wide and slants up from ground to bottom hive entrance, giving bees a larger landing deck. I pack the space inside the stand with fiberglass insulation or steel wool to keep mice from gnawing in during winter and ruining the hive and the bee colony.*

Hive Base and Bottom Board: *A solid wood base with wood strips nailed around three sides. Hive sits up on the strips and the open side provides the hive entrance.*

Hive Body: *Four-sided boxes with handle holes milled out of each side. Metal rails nailed on inside for frames to rest on. We have two full-size [9½ inch deep] bodies for the brood area and two shallow [6 inch deep] bodies to collect the honey in. The latter are called "supers", also come in full-size, which weigh fifty pounds or more when full of honey and that's too much to lug around.*

Frames: *Two precisely milled side bars lock into the top bar and a split bottom bar goes on last. All are glued, nailed or both to form a remarkably strong open frame.*

Comb Foundation *goes into the frame. Thin sheets of pure beeswax embossed on each side with the design of the cells the bees will build, or draw, as they use the foundation wax as well as wax they secrete themselves. The foundation sheet is slipped through the split bottom bar of the frame and attached to the top bar with a thin wooden "wedge" that is nailed in place. Wax foundation alone is not strong enough to hold repeated loads of brood without sagging badly so several strengthening techniques are used. Big apiaries wire the foundation on with expensive machines and the manufacturers sell foundation laid on plastic sheets. I don't like plastics, and I've heard that bees don't either. We use foundation with several vertical wires factory-imbeded in the wax that is held up at the sides by four pins—miniature clothespin-like things that go through holes in the side bars and hold the foundation in their "jaws".*

Inner Cover *fits over top of the hive to keep warmth and moisture in. Has a hole in center which will hold a "bee-escape" used to clear bees out of a honey-filled super. Usually the hole is kept sealed.*

Telescoping Cover *tops it all off. A solid lid a bit larger than the hive body with side boards that fit down over the top couple of inches. A sheet metal top provides needed protection from rain.*

the snow-white sealing wax off the honey comb. The uncapped frames then go into a hundred dollar or more extractor which whirls them around, removing honey by centrifugal force. The emptied frames, with cells intact, are put back in the hive for a refill.

WHAT YOU DON'T NEED

Equally unnecessary for the homesteader, I feel, are section supers where bees pack their honey into little square wood frames that pack nicely for sale with comb and all. The main disadvantage is that the bees don't like all the frames posing barriers to easy travel in the hive and they tend to swarm—when the bees get to feeling crowded and three quarters of the colony takes off with the queen for a new home. For home operation, I think it is easiest just to cut the whole comb, bought foundation and all, from a frame. What we don't eat comb and all is heated to 170° in the stove. This destroys the vitamins but is necessary to pasteurize the honey so it won't sour in storage. It also melts the wax which is poured off the top and turned into the finest hand-dipped candles known to man. (Melt wax in a tall can and just dip in lengths of wick. Regular string won't do; we use wicking. The best comes with a burnable wire core so it dips easily. To stiffen non-cored wick for hand dipping you have to swish a length in wax, then stretch it tight till it cools and hardens. Wax melts at a low enough temperature so that it won't burn your hands, though it does smart a bit.)

Arrival of the Bees

OUR FIRST BEES

Our first hive arrived and was built a week before the bees were scheduled to arrive from Georgia. When the postmaster called and said the bees had come in I could hear them buzzing over the telephone, and I want you to know that is a scary sound the first time you hear it. Fortunately I had obtained the Root Company's free brochure on handling package bees because there was no instruction sheet on the shipping cage, a wood frame box with window-screen sides. There was supposed to be a can of sugar water hanging from the cage top and a mated queen in a separate cage beside that. I couldn't see any of it, because everything was covered by a solid mass of bees, each with a stinger and humming ominously.

Once home I donned gloves and mosquito net, buttoned

up collar and sleeves, tucked my trouser tops into my boots and carried the cage up to the hive. After removing the hive tops and half the frames, I pried off the wooden top nailed to the mailing cage. Under it was another board with a key-hole shaped cutout. In the circular part was the top of the syrup can while the slot part held the queen cage top with a bent nail sticking out. Hands shaking, I flipped the cage over, shook gently till the top of the syrup came out an inch or so, then turned the cage right side up again. As I pulled the can out slowly the bees were scraped off and fell buzzing to the cage bottom.

HANDLING BEES

Removing the syrup can

Fully expecting the entire swarm to issue forth, stingers at ready, I pulled the can all the way out. Nothing happened. I scraped the few bees still clinging to the can bottom back into the cage and still nothing happened. Relieved, I examined the queen cage. Thinking the bent nail held it in place, I pried it out with my knife. I'd been right, I realized, as the queen cage dropped into the cage to be swallowed up by the bees erupting in a brief but mean-sounding growl. (The queen cage should have been slipped out, nail still in, out into the circular part of the cutout.)

Removing the queen cage

Petrified as I was at that moment, my mistake proved to be a lucky one that taught me not to be afraid of the bees. I had to go fishing in them to extract the queen cage, which must be hung between two frames and the colony left undistrubed as the workers get used to the queen and release her by gnawing out a candy plug in the bottom of her temporary prison. I tried fishing with a pair of sticks but only succeeded in causing an angry buzzing. Finally, working *very, very* gingerly, I pried loose the screen from one side of the shipping cage and stuck in my gloved hand. Not a bee complained. With mounting confidence I felt for the queen cage and extracted it. Through its screen windows I could see the queen, about half again as long as the half-dozen ''nurse'' bees sent along to keep her company.

Really confident now, I removed another frame from the hive, suspended the queen cage from the adjoining frame by its nail, and put the package containing the rest of the bees in the space made earlier by removing five frames. The hive tops went back on and I inserted the entrance reducer and feeder to supply the food an established colony would obtain from stored honey. The reducer is a length of wood the size of the hive entrance with a large and a small notch cut in it. The feeder is a wood block with a round hole drilled halfway into the top and a beespace-sized channel reamed through from the round hold out to one edge. Into the top hole fits a mason jar lid with feeding holes punched in it.

**FEEDING
SOLUTION**

I had a jar filled with feeding solution of sugar water which I inverted, lodged in the feeder, then inserted the feeder into the large notch in the entrance reducer. This left a bee entrance of less than an inch—essential with a starting hive so the relatively small colony can defend itself. At first I tried feeding maple sugar, dissolved in water, at the rate of five pounds of sugar to one quart of water. The bees didn't like it. They were happier with the same formula but using our semi-refined sugar. Most acceptable was to make invert sugar, which you can think of as partially digested sugar, and which is found in honey. I dissolved five pounds of white sugar in six cups of hot water, added a tablespoon of fresh lemon juice, and let it simmer for at least a half hour. The bees literally eat the stuff up at the rate of a quart or so a day. Feeding, by the way, is kept up with a new colony till they move on to natural nectar in flowers. You know they've switched when their consumption of sugar water diminishes greatly.

A Long Week

With the hive set up we commenced the long week's wait during which the colony must not be disturbed for fear of the worker bees killing the queen. I spent an inordinate amount of those seven days sitting on the orchard knoll watching the newest members of our organic family buzz in and out. Soon I lost any remaining fear, learning that so long as I moved slowly and made no sudden gestures, I could join in bee life without being bothered. It is fascinating to watch the bumbling, stingless drones roar around importantly, to see workers buzzing in and out, to watch a guard bee chase away an ant with the temerity to try stealing some honey, or to observe the life and death battle between several guard bees and a white-faced hornet searching for caterpillars to feed her developing young.

**OPENING
THE HIVE**

When the week finally ended I fired up the smoker with kerosene-soaked rags, but feeling like part of the bee family, I disdained face net or gloves. According to instructions, I puffed smoke into the hive entrance, then raised the covers and smoked the tops of the frames. A bee sailed out and landed on my cheek. I swatted it but not before it stung. The sudden movement brought a dozen more bees out after me, and I took off down the hill trying to remove the stinger in the approved fashion, scraping it off with a finger nail instead of pulling it between two fingers, which just squeezes in more venom.

What did I do wrong? I hadn't made any quick movements. The day was sunny and warm, not the cool or rainy weather that makes bees grumpy. Though there was no nectar coming in yet to distract the bees completely, they were being fed. That only left the smoker and I've never used it since.

After giving the bees a few minutes to get over being smogged, I tried again. The colony let out a low hum as I removed the inner cover again. I leaned it up against the hive, inner face down so the several hundred bees clinging to it would experience less of a sudden light change. Moving very slowly, I reached in and removed the queen cage, now empty. Next to come out was the shipping cage. I had to use the knife to cut the cage away since the bees had built several tiers of snow-white comb between the cage and hive wall. And to my delight, the comb was filled with honey—undoubtedly made from the sugar syrup.

Still moving slowly, I removed the frames which the bees had been working on for a week. Most of each frame was covered with workers busily drawing comb, but on one I found what I was looking for, a sheet of fully-drawn comb, each cell with a tiny pearl-white egg affixed to its bottom. On the opposite side of that frame were more cells containing young in the second or grub stage of growth, looking like small versions of the cutworm grubs that cause garden problems.

THE INSPECTION

The presence of eggs and grubs was a pretty good indication that the queen had been accepted by the colony and was busy at her only job, laying several thousand eggs a day. I wanted to be 100 percent sure, however. If she was not present I could have a replacement mailed up from the South in just a few days. With live eggs in the nest the workers would have created their own queen replacement by building the downward-hanging, peanut-shaped queen cells and feeding up a worker grub to royal status with special "royal jelly" food. This process would have taken weeks, however, the weeks needed to build up a strong colony for the coming honey flow.

The third time through the frames I finally spotted the queen. She's much longer than the workers and has a less distinct coloration than the Italian workers with their three yellow abdomenal bands. Still, it takes practice to pick the queen out. Assured that she was alive and laying well, I replaced all the frames, recapped the hive, and took the cages and the sections of comb I'd cut out down to the house. There we all had our first taste of the delights to come later in the year when we would harvest from each super almost fifty pounds of nature's most

Getting the hive out of bed after the winter: Straw packed into a shallow super on top is pulled out, then bales stacked up in back and at the sides are removed.

healthful sweetener and enough wax to provide a goodly portion of the year's supply of dinner candles.

A similar procedure is followed each spring now that we have several established colonies. Toward the end of March I remove the hay insulation from around the hives and tear them down. If the queen is alive and laying well all I have to do is make sure she has laying space. If more than ten of the twenty frames are still filled with honey, the excess is replaced with empty drawn frames or foundation. Conversely, if the colony is very low on honey I will feed them till apple blossom time.

If a colony has been winterkilled or has had too few flying days in the winter so it develops dysentery and dies early in the spring, it can be replaced with package bees like those used to get started. Perhaps better, and certainly cheaper, is to remove the dead bees, seal the hive, and at the height of honey flow (and peak swarming time) in June, divide a strong colony. This will be explained when we get to the orchards in a later chapter.

In any event, we try to be finished with the bees and have all the sugaring equipment cleaned and stored by April 1 because it's time to get gardening.

4

Gardening
Before the Bugs

As the snow finally melts and mud time comes on us, the maple flow dribbles to a stop, bees turn their attention to raising brood, and we can look to the gardens. On April Fool's Day we are six weeks before our last frost of the season plus another two insurance weeks before we can set out frost-tender plants. So, many folks might think us fools indeed to spend April in gardening high gear. We have persuasive reasons though, some of which are shown on the chart on the facing page.

There is nothing scientific about these curves, they are from pure experience and simply an attempt to put on paper the most important weather variables: temperature, sun, and day length. I'd have included some indication of rainfall too if it were predictable here, but it isn't. The chart plots curves which depict the garden value of sun and temperature against a time line running from March to October. If nothing else, it shows why there are so many abandoned farms in upper New England. Sun is strong enough to grow crops from mid-May to mid-August. But temperatures are good from early June to the end of September. That gives us a prime growing season with good sun and temperature of only seventy-five or eighty days, from June 1 to mid-August. After August 20 or so is ripening time, but growth has almost stopped. (This information is good only for our own hillside but it may give someone else a way to get a handle on their own growing conditions. Hope they are better than ours.)

NEW ENGLAND'S
BLACK FLIES

The chart also suggests several million other reasons to really go at it in April—bugs. I suppose every section of the country has its own flying, biting reasons to appreciate the genius who invented the screened-in porch, but the woods of northern New England sport a special four-species line-up. In early May the first black flies arrive, to peak toward the end of the month in great clouds of tiny gnats that buzz into your eyes and nose and crawl under your socks to leave itching welts. If the spring is dry they only last two weeks, but then come the mosquitoes, then the no-see-ums, and finally the deer flies which can extend insect miseries well into July. We've been asked why we chose such a buggy place to take up homesteading and I must admit I have asked myself the same thing many times each spring as I go up to the gardens to see the bugs rise to greet me in hungry, humming swarms.

Often people ask why we don't do the *intelligent* thing and use insecticides.

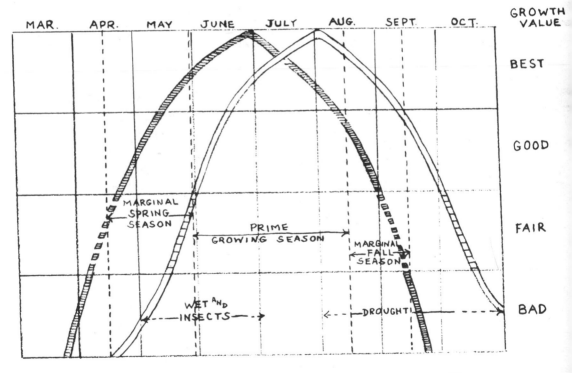

SUN – ▤
TEMPERATURE – ▢

Rachel Carson Was Right

DON'T FOG

When we moved to the homestead we had only partially accepted the warning of the dangers of chemical pesticides first publicized in Miss Carson's landmark book, *Silent Spring*. We had long since stopped using any bug sprays on the food plants, but we did have a yard fogger. It was the kind that attaches to a lawnmower exhaust and turns oil-based insecticide to a billowing fog. When we found our barn filled with fat gray spiders whose webs filled every doorway, wetly plastering our faces each morning, I ran the fogger through the barn just once, filling the stalls and loft with fog. Next morning there were no bothersome webs.

But other effects began to show up immediately. First was a brood of barn swallows, fully feathered and just a few days from leaving the mud nest located on a beam just under the loft floor. The second web-less morning the nest was quiet, the parents flying aimlessly around outside. While inspecting the dead nestlings I found a half-dozen dead bats in the loft—fallen from their daytime roost in the rafters. Several were females with tiny young clutching their fur. And the next day a hen died—none of the normal symptoms, she just keeled over dead in the henhouse which is located at one end of the barn.

Coincidence perhaps? I don't think so. Too many creatures died too close together of inexplicable causes. The birds, bats, and hen must have been exposed to too much bug killer as it accumulated and stayed in their roosting areas and perhaps their food. I suppose I could have had autopsies performed on the victims for final proof, but circumstantial evidence was good enough for us. The fogger went to the dump.

Then the flies began to increase in number. Any homesteader with animals will have some flies no matter how conscientiously he composts manure and cleans quarters. But our fly population, free of the natural control posed by hundreds of hungry spiders, exploded to the point that sitting in the back yard was impossible on any warm, sunny afternoon, for all the buzzing, diving flies of every size and description.

NATURAL SYSTEMS

It took the spiders three years to build back up to their prefogging population; fortunately the insecticide in the fog wasn't long lasting. The bats and swallows (all bug-eaters) were quickly back to their former strength too, and with each succeeding year the fly problem got better. Now we take pains to avoid the morning's webs, not because they are a minor annoyance when run into but because they are a major fly-control measure. To be sure, they get help—fly screens over the animal pen and house windows, quick composting or covering of fresh manure, a half-wild flock of bantam chickens that scratch up fly larvae, plus stringent sanitation measures. But no poisons are used. They are simply not needed so long as we keep nature's self-balanced life system intact.

This is the essence of our concept of organic living—doing everything within reason to make sure nature can operate the world unencumbered. Where we violate natural law, such as having an unnaturally dense population of livestock in a comparatively small area, we must provide some compensation, such as covering manure, so that natural equilibrium will be reestablished. When it is, spiders will keep fly population at a normal level. So much for lectures. Up to the garden!

Toward Self-Sufficiency

A major element in the organic life is production of most of our food. (We'll never grow it all; we like orange juice, and citrus trees just don't grow in New England.) But to grow a year's supply of most fruits and vegetables in only seventy-five days of prime growing season with some forty more days of marginal weather, plus bugs, takes a lot of preliminary work, much of it in the hotbed/cold frame.

HOTBED
COLD FRAME

We find that a small frame, only five feet wide by three feet deep, is satisfactory. It is located on the south side of the house where it gets sun from about a half-hour after sunrise till noon, when it is shaded by the barn. That gives only seven hours of sun in April and May, which is barely adequate but all that can be expected in a clearing in the woods.

The frame is a simple affair, made from scrap wood. The sides are shaped like a sixty-degree triangle with one point chopped off. The high edge, where the sides butt up against the house, is a foot and a half tall and the sides slope forward and down to a front edge that is some six inches above ground level. These sides aren't

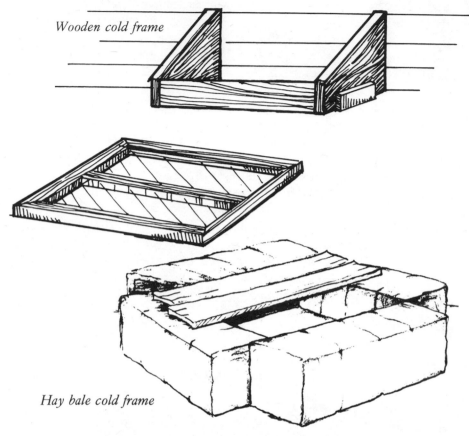

Wooden cold frame

Hay bale cold frame

attached to the house, but are kept in place by stakes nailed on and pounded into the ground. A seven-inch by five-foot board is nailed across the front. For a cover I use two old window shutters with the slats knocked out, and the frames are covered with a seimiflexible plastic sheeting with wire imbedded in it. These are laid crosswise over the frame, the bottom one resting against the inch of front board that extends up beyond the sides.

This frame isn't airtight, but the only problem we've had was one year when a freak cold snap with high winds burnt a few tomato plants. After that I stuffed hay into the cracks between the frame and the house clapboards and had no more difficulty. I know a lot of folks make cold frames from cement blocks and a window sash with hinges and all, but I see no reason to spend the effort and money. As a matter of fact, for the first few years all I used was old hay, one bale front and back and two for the sides, with boards laid over the top at night. It worked adequately, and still does; I rig such a frame around a double row of kale late each fall and we enjoy fresh greens into January or February.

The one problem with an old hay cold frame is that the bales rapidly soak up water and begin to rot, providing sanctuary for all manner of plant-eating pests. When I finally broke down and built the "fancy" frame it was mainly to dispose of the insect problem, but also to be able to use it as a hotbed early in spring. The first year I followed the usual practice of burying electric heating cable deep in the frame's six inches of loam, and grew plants directly in the frame. The problem was, I couldn't get rid of slugs, which gnawed on the seedlings no matter how many plates of stale beer were set out. The little gray pillbugs or rolly bugs abounded under the hay, to come out at night and feast on tomato roots and leaves. There was even a hotbed mouse that burrowed into what must have seemed mouse heaven—an electrically heated apartment with three free meals a day.

INSECT TROUBLES

The final solution to all these problems took a year of experimenting and some digging, but it works. First I dug out all the loam and topsoil down to about eight inches under the frame. The sides of the hole were lined with mouse-proofing, aluminum roof flashing nailed to the inside of the frame sides and front. In the bottom of the hole I put three inches of sand. On top of that went a good inch of rock salt, and then another three inches of sand. The heating cable was snaked out on the sand in a continuous series of "S" curves. Another two inches of sand went on and I had a mouse- and bug-proof frame. The rock salt is the secret, acting as a barrier against any critter that might want to burrow up into the frame from the soil beneath. Even a bug that got in on the deep flats used to hold the plants would likely be killed if it dug down into the sand very far.

A SOLUTION

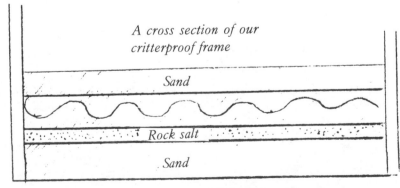

A cross section of our critterproof frame

Sand

Sand with heating cable

Rock salt

Sand

Firing Up the Frame

Each year on April 1, I plug in the heating cable and fire up the hotbed/cold frame. Most years the frame is buried in old snow and I often have to chip the shutter covers free of ice. In a few hours the cable has heated the sand enough that I can grub out any ice chunks, and by the next morning the frame is ready to receive its first plants of the year.

WOODEN FLATS

When the cold frame was being constructed, I also nailed together several deep wooden flats or planting trays from scrap lumber. Measuring one by two feet by four inches deep, they fit perfectly on the shelves of our window greenhouse and six of them fill the cold frame with room to spare for an odd flowerpot or two.

The flats' dimensions were also decided with an eye to using three-inch-square peat pots or the six-pot "jiffystrips." Each flat will hold three strips, or a dozen and a half pots. Our main reason for using the peat pots, as against growing plants in a single block of soil, is to restrict the root growth of each plant to its own growing medium so that no roots are lost in transplanting. About the time the frame is fired up we transplant all started seedlings from their egg-carton nurseries to peat pots. The cold-tolerant members of the cabbage family go directly to the hotbed, there to harden off till it is time for them to go into the garden, at which time their place is taken by the tomatoes and other warm-weather-loving plants.

SEEDING IN THE FLATS

We always seed directly into one flat. I place several layers of newspapers over the slat bottom and fill to the top with starting medium. With a kitchen knife, Louise plows furrows about a half inch deep and two inches apart across the narrow dimension of the flat, giving some dozen rows. She plants two rows each of an early, transplanting variety of beet, spinach, and kale, and three types of lettuce—our favorite **Buttercrunch**, an improved Bibb-type that takes a good two months to mature, the month-and-a-half loose leaf variety **Black Seeded Simpson**, plus an experimental variety, most recently Stokes's **Premier Great Lakes**, a three-month-heading lettuce. To date we have found **Little Egypt** the best of several early beet varieties we've tried. It matures a small but sweet bulb and a moderate amount of greens about six weeks from planting. Our favorite spinach is Harris's **Winter Bloomsdale**, and we find **Vates** (blue, curled) more attractive than the yellowish Scotch varieties of kale.

Frost danger past, Louise removes young flower plants from the cold frame for planting up in the garden.

The direct-seeded flat is placed in the hotbed and kept well watered. Over several weeks of alternate snow, rain, blizzards, frost, and generally wet April weather, the cool-germinating seed will sprout happily and develop into stout little plants eager for transplanting.

Earliest Spring Planting

It usually happens that the day our garden soil is dry enough for early spring planting coincides with opening day of trout season. Having a maximum amount of time and energy for practicing the angler's science over the early spring weeks is one

**LONG REACH
SQUARE**

reason I plant in blocs by what I call the Long Reach Square system. Many of the cold-tolerant vegetables have tiny seeds, tinier seedlings, and a relatively long germination period. However, in mid-spring as each miniscule carrot and lettuce seedling is working away to grow big enough to be seen without a magnifying glass, there are a million weed seeds sprouting too, and many of them can outdistance a garden plant in just a few days. With the Long Reach Square system (let's call it LRS for short), we can plant and weed quickly, thoroughly, and sitting down, which can be a blessing to a winter-softened back. Probably more important, the process lets us avoid stepping all over the still damp soil, compacting it to the detriment of plant growth.

An LRS planting is nothing but a series of blocs, like squares on a checkerboard, with narrow paths between them. The size of each bloc should be twice the distance the gardener can reach by stretching long, but without straining, while in a comfortable sitting position. For me that's three feet, so my

Early spring planting: Using the planting board to seed the "Long-reach Squares".

squares are six feet on a side. I suppose you could make the square
a rectangle—a long, narrow planting bloc—but that would require
a lot of stretching. With the square, only the very center requires
the real long reach.

I figure the productivity of the soil in each square has to be
increased by about half to make up for the land area lost to all the
surrounding paths. Each thirty-six square feet of planting space
requires twelve linear feet or twenty-four square feet of two-foot
wide path, or about 40 percent of the garden devoted to path. The
increase in productivity is obtained by reducing normal planting
distances, successive cropping, and digging in a strong dose of
plant nutrients after many crops are harvested. And then, while
they aren't producing food, the paths are not at all waste land, but
are lying fallow. We let the weeds grow happily in the paths,
pulling out only the pernicious perennial grasses, but tilling in the
weeds several times a year before the weed crop can go to seed.
Weeds and culls from the squares too are left in the paths to decay,
so that by year's end the soil under the paths has actually increased
in organic content and plant nutrient value.

LOOSEN THE SOIL

We try to make the actual planting process as easy on
ourselves as possible. The first and most important part of the job
has already been done: a thorough tilling of the garden plots the
fall before. Now, after the alternate freezing and thawing of early
winter and spring, the land needs only a surface pulverizing job to
be ready to accept seed. The top dressing is done with a gadget
forest fire jumpers know as the McLeod fire tool. It is a
combination hoe and rake: a nine-inch-square tool-steel blade, one
edge flat and honed razor sharp for hoeing, the other cut into six
heavy rake tines. Its major advantage is weight. Unlike regular
garden tools which weigh a piddling pound or two, this one tops
five pounds. That means all I have to do is haul it around over the
soil; its weight does the work. Even when I have to heft it to break
a particularly stubborn clod, it's easier than with a regular hoe or
rake. With the lightweights you have to work going up and down.
With the fire tool I do the raising part, but the five-pound head
does the hoeing.

McLeod fire tool

LAY OUT PLOTS

Once the soil is loosened on top, I lay out the LR Squares
using eight-foot lengths of poplar sapling cut from the stand
bordering the old pond site up beyond the west wall. The saplings
are quicker and easier than laying out the plots with string, and
later in the year I'll use them to support the pole beans. To make
planting furrows I use a homemade planting board. It is a six-foot

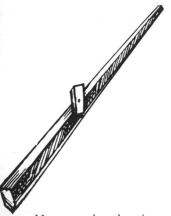

length of one-by-three-inch scrap lumber with one edge planed to a knife edge and a six-inch-long handle nailed on so it slants forward about one-third the way along the board's length. Holding the board with the handle pointing away, to give better leverage, I slide it back and forth and wiggle it from side to side to make a trench of most any depth and width. After planting seed, the board is used to scrape soil on and then to tamp it down. All this can be accomplished in an easy sitting position if it's a lazy day. The planting rows, by the way, go in a north-south direction. Lined up east-west, the way the sun moves, plants in the southerly rows would shade the northerly ones, especially early and late in the season when the sun is low in the sky.

Homemade planting board

Vegetables as They Go In

Most books you read list vegetables in alphabetical order. Nature doesn't list things in alphabetical order, though, and neither will I. We just naturally think of vegetables in the order they are planted, so that's how they will be discussed. I'll only list the ones we plant a lot of and eat regularly, too. We always have a few throw-away rows of purple-podded beans, celeriac or whatever, and despite years of failure I haven't given up on lima or soya beans. But that's for fun or experimentation.

Parsnips

Nothing on earth tastes quite like fresh-dug parsnips deep-fried in peanut oil, lightly salted, and served with a fresh pork roast and steamed broccoli. Despite its cold tolerance, getting this slow vegetable with a germination period of up to a month, to mature by fall requires the earliest of early spring planting. We grow the Stokes **Hollow Crown Improved**, which is about half as long as most varieties and well suited to shallow and rocky New England soil. Since the roots often stay in the soil for a full twelve months, I plant some thirty feet of row (a small seed packet's worth) in one or two LR Squares at one or another end of a garden plot where they won't interfere with fall tilling. The seed, together with a scattering of radish seed, is planted in rows two feet apart in very shallow trenches that I fill with fine sand. Even the best organic loam can compact over the parsnip's long germinating period to the point that the weak-kneed little seedlings can't get through. The sand solves this problem.

The young radish seedlings help mark the rows, and are ready to go into the relish tray just about the time the parsnips come up. Through April and early May, the parsnips get a weekly inter-row hand-weeding and are gradually thinned to four inches apart. Weeding is accomplished with a dandy little tool that looks as if a six-inch length of hacksaw blade has been bent in a "U" shape and riveted to a stout wire handle. They come pointed or square and are sold for eighty-five cents by D.V. Burrell Seed Growers Company, Rocky Ford, Colorado 81067. Called honestly enough, garden weeders, they are sold to home gardeners only by Burrell so far as I know, though they are used in the thousands by large commercial truck farmers.

A ROW MARKER

Garden weeder

In mulching and bug-control, we treat the parsnips the same as our much larger planting of carrots. We used to dig parsnips in the fall and store them in moist sand in the cold cellar, the same as carrots. Of late, though, I've been companion planting them with the double row of kale that's wintered over in the old-hay cold frame with boards over the top. Both vegetables stay in the ground all winter, the parsnips improving with age to taste supersweet when dug in spring. Incidentally, they tell me that if you let overwintered parsnips grow up and go to seed the second year, the leaves are highly poisonous though the root never is. None have ever lasted that long at our place, so I can't vouch for it.

Parsley

The runner-up for lengthy germination time is parsley, though sprouting is aided if seed is soaked overnight in cold water. We use this vitamin-rich vegetable for much more than a pretty garnish. Fresh during the growing season or dried in winter, it goes into most stews or is sprinkled on soups. I nibble the stuff every time I pass a row just because I like the perky taste and I hear it has more natural vitamin C per ounce than any tablets.

We grow two kinds of parsley. Sometimes in an LRS row or two, sometimes alternating with the bug-repelling marigolds and nasturtiums, we grow the European favorite variety, **Bravour**. The plants bear heavy crops of deep green, well-curled leaves of a rich but not bitter taste. The seed is available only from Stokes, I believe. The other variety is **Japanese Perennial** with a flat, light-green leaf and an unusual flavor. The yield isn't too heavy, but it is planted in Louise's herb garden which could use a bit more sun. Another unusual seed, it is available only from Nichols Garden Nursery in Albany, Oregon, so far as I know.

Sugar Peas

The last member of our Polar Bear Club is sugar or snow peas, the kind you eat pod and all before the peas themselves have developed. The only places you can buy them are in Chinese restaurants where they add their special sweet-crispness to many favorite dishes, and in a few specialty food stores where they go for a small fortune. We plant the small, dark, smooth seed in a single row, two inches deep and not quite touching along the back picket fence. We order the rankest and tallest-growing kind available. The vines grow four or five feet high and the blooms are a highly ornamental dark lavender. We pick the abundant crops just about the time Louise's morning glory vines are beginning to take over the fence.

POOR STORERS　　　The only problem with sugar peas is that they don't stand up to freezing or canning—they lose crispness. So we have to eat them as they come, raw in salads, boiled briefly and served with butter, or sauteed in Oriental dishes. Years when the crop is unusually heavy, special friends receive a gift that truly is unusual—a quart of a wonderful vegetable that a lot of people have never even sampled.

Beets

Toward the end of April, when the threat of a really hard frost is practically nil, we put in the beets, both transplants started earlier in the month in the hotbed, and seed. Each seed of most beet varieties will produce two or more separate plants which will have their roots tangled together. I dig the **Little Egypt** seedlings from the nursery frame and put the roots immediately into a

Would you believe planting in a foot of snow? But that's the only way to get in the snow peas we eat pod and all—plant them under the mulch of old hay I am applying here.

shallow pan full of cool water. The planting board is used to make furrows a foot apart and worked to a depth of two inches. I separate the beetlings by swishing them gently in the water, then plant them just three inches apart and firm the soil very firmly around the roots. Usually there aren't enough transplants to fill more than a row or two in an LR Square, so I complete the square with seed of the early variety.

Two or three weeks later, between each row of early beets I plant half-inch deep furrows of **Long Season** or **Winter Keeper** beets. A very late variety, **Long Season** grows slowly to a rough, elongated shape that is not as pretty as globe beets. However, the root is the sweetest of any beet we have grown. Equally important,

SUCCESSION PLANTING

if mulched they will retain their superb quality through the hottest summer and fall weather. The greens are coarse looking but tasty as any and they are so frost tolerant that we often leave them in the ground through October. By that time the roots are large enough so that just one or two will provide a full meal serving.

We enjoy beets, roots and greens served together, perhaps twice a week during the growing season. Fall-harvested beet greens go to the livestock unless the spinach crop was a failure, in which case they are frozen. The roots go into moist sand in the cold cellar and provide perhaps one meal a week over winter. Planted as indicated, a single LR Square provides our year's beet requirements. The first meals of beet greens come in May from thinnings of the transplants. Next are thinnings of the direct-seeded early variety, then mature early beets pulled while smallish before they overpower the small late beet seedlings growing just six inches away. Midsummer meals come from **Long Season** thinnings and fall and winter servings from the mature late variety.

FERTILIZING   We have never had any serious disease or bug problems with beets. The annual liming of the plot prevents yellows and our heavy fall application of animal manure fertilizers provides the trace of Boron needed for strong, black-rot-free bulbs and the nitrogen for luxuriant leaf growth. When the early crop is harvested I mulch the late planting with six inches of old straw bedding from the goat pen. The mulch will keep in moisture and rain washing down through it will carry an extra dose of natural plant food in the form of a dilute ''tea.''

Lettuce

Just about on a par with beets, so far as cold tolerance goes, is lettuce. The small plants started in Louise's nursery flat are set into an LR Square in rows a foot apart. The leaf lettuce is spaced two plants per foot, the Bibb-type eight inches apart, the heading variety a good foot apart. Empty rows are filled with more seed of the Bibb-type, **Buttercrunch**, scattered thinly in half-inch-deep rows and firmed extra well. In a good spring we will have our first salads from thinnings of the leaf-type transplants at the end of May. The plot will keep producing till the last heading-type plant comes out in July, when a post-bug season planting will be coming on.

We've never had any avoidable problems with lettuce. Slugs are said to be a nuisance, especially later in the summer when deep mulches are necessary to conserve soil moisture and keep the plants' feet cool so they won't bolt to seed, that is, the leaves turn bitter and the plant puts out a tall flowering seed stalk. Our gardens are literally hopping with toads and leopard frogs who feed on the same nocturnal schedule as the slugs so that the slimy little pests never have a chance to get established. Mulch, by the way, is the best way to attract these bug-eating amphibians. It gives them a place to hide out of the day's sun and heat. In spring before mulching time I try to leave little heaps of old hay around, and sure enough after the weather warms every odd one will be sheltering a frog or toad on any given day.

Toad

Spinach

From what you hear about spinach's reputation we must be a bit peculiar—kids too—but we all love spinach served raw in salads, steamed as a hot dish, creamed over hardboiled just-gathered eggs on homemade bread toasted for lunch, as the basis of a souffle, or any way we can get it.

Our current favorite, **Winter Bloomsdale**, is a variety with leaves that are fat, thick, and deep green. The flavor is pure spinach, and the greens a veritable vitamin and mineral supplement with high content of calcium, iron, and Vitamin A. It also stands well, and will resist bolting well into the warmth of late spring and early summer.

Most years I put in two LR Squares of spinach, using a full ounce of seed to plant about seventy-two feet of row. Transplants from the nursery flat are spaced six inches apart in rows a foot and a half apart. The large, wrinkled seed, soaked overnight, is scattered about four to the foot in the same row spacing, under a scant half-inch of soil. Plants as they grow are thinned to stand two per foot, the thinnings providing several mid-spring meals. The mature plants are pulled and quick-frozen no later than fifty days after planting, the middle or end of May. With good weather these two LR Squares provide a month of salad greens and fresh spinach, plus one or more weekly servings the year round, plus several fine meals of stalks and cull leaves for the animals.

Spinach isn't heir to any problems we can do anything about. During an occasional wet and sunless spring the plants will

OUR FAVORITE

PLANTING

be stunted, producing perhaps a half crop. When this happens we make up for the loss by freezing the fall harvest of beet greens that would otherwise go to the livestock.

Sweet Peas

We call green peas "the converter" because we've seen more people turned on to gardening by fresh peas than any other vegetable. Freshly picked, the green pea has a sweetness and heady flavor that bears no resemblance to what is sold fresh, frozen, or canned in any store. Our own year-old frozen peas are better than servings we've had in any restaurant simply because ours were put up when less than a half-hour old. It will come as no surprise that we adore green peas and grow 200 feet of row each year.

The soil likes peas also. A legume, the plant takes free nitrogen from the air and with the aid of naturally-occurring soil bacteria "fixes" it into a form plants can use as food. And on top of that, the vines and shelled-out pods are used to mulch following crops and when tilled in add masses of organic fiber to the soil. What a vegetable! Superb food, nitrogen fixer, mulch, and green manure all in one.

Over the years we've tried most every kind of pea variety and cultural method. In our book, the new early varieties are inferior in quality to the older kinds, and they are earlier by only a few days. We don't like the new dwarf vines either. They don't produce much, and picking them means getting down on your knees and scrabbling around in a lot of clingy vines. So we've gone back to the old ways.

PLANTING   First I use the hoe to grub out 200 feet of north-south row across one of the fifteen-foot-wide strip gardens. (We'll get into the reasons for them later on.) For peas I use double rows about a foot apart laid out along pairs of the same poles used to set out the LR Squares. Furrows are a couple of inches deep and each pair is a good yard from the next. Into each I scatter an equal mixture of seed of **Lincoln** and **Tall Telephone** or **Alderman** seed. It takes a pound of each. I try to get about two seeds to the inch. This is wider spacing than the "close but not touching" usually recommended for peas. But these are tall-growing, vigorous and hungry varieties and they produce better given a bit more planting space.

Next step is to put up supports. Traditionally, brush or SUPPORTS
birch saplings are used to support peas. After tiring of hauling half
a wood lot of brush around each spring, I made supports of
five-foot high chicken wire. It is cut in fifteen-foot lengths with a
seven-foot stout pole wired on at each end and in the middle. The
poles are hammered in the ground so the wire cuts the middle of
each double planting row and the end poles are guyed from the top
with lengths of wire fastened to ground stakes. The guying system
is fairly complicated, as the diagram shows. This setup
will keep the wires up in a high wind even when the vines are full
grown and I can get between the rows to cultivate with the tiller.

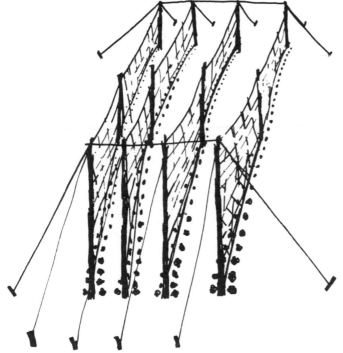

Pea supports and guying setup

The peas germinate together and flower a week apart. In
nine to eleven weeks from planting the **Lincoln** turns out two good
pickings of medium-sized pods and peas on vines that never get
higher than three feet. The **Alderman** plants grow wildly, often
putting out a foot or more of waving vine above the top of the wire.
The **Lincoln** vines are beginning to wither when the **Alderman**
comes on with huge crops that require daily picking for up to two
weeks. Both varieties are superb eating, though the tall variety
produces a large pea that gets mealy if left on the vine too long.

PEA ROTATION

Rotation is the key to sweet pea success and we never plant them on the same plot twice in five years. Poor weather can reduce crops, but there is nothing to be done about that. I cultivate between rows once a week through June to keep down weeds and pull any large weeds growing right in the planting row. The lush vines themselves do a fair job of shading and mulching their own roots, but if weather warms and the soil begins to dry before the pods set, I might throw a chopped hay mulch along the rows.

Once the harvest is in, but while most of the vines are still green, we go along the netting, pulling the vines up and off. The supports are removed and stored in the loft, and after the late broccoli, cabbage, and cauliflower are set out where the peas had been, the old vines are pulled up to them for a wet mulch. So in their ancient age the pea plants serve a new gardening purpose. As I said earlier, what a vegetable! About the time I get the peas in, ice goes out of this pond I know about. It's a hike back in the woods, but the little native brook trout are right on shore, just waiting for a sore-backed homesteader to get the fly rod disentangled from the overhead branches and lay half a red wiggler on a tiny barbless hook right on their hungry noses. If any dry fly purists are rankled by this dishonorable act, all I can say is there aren't any hatches that early. Besides, I only have a few hours. There's another half a garden to be planted. But first is that initial trout fry of the year—the little fish gutted only and fried in hot butter, served with the last of the wintered-over kale and real potatoes with butter and parsley flakes. The potatoes are another story, so we'd best get back up to the garden.

Potatoes

The plant breeders haven't managed to ruin our late baking potatoes. An Idaho is still an Idaho. But give the boys a chance; they've sure done a job on boiling spuds. The problem is, folks these days have forgotten how a potato tastes. For one thing, most shoppers get mashing potatoes in a box of instant flakes that taste to me like they contain a dose of kerosene. And when you do find anything but expensive bakers in a store, it's something with a texture like wax and a flavor to match. They have down-Maine sounding names on them, but that is about the only resemblance they bear to real potatoes.

At our place early potato means just one thing: **White Cobbler**. This is the good old Irish potato that is dry and mealy, with steam that makes your mouth water when you pop one open. And they taste like something, which a potato ought to. But try to find them in stores! The plant produces tubers that today's shopper would find ugly, typically one huge nobbly spud, a half-dozen medium to small tubers in odd shapes, and a good handful of half-dollar to penny-sized miniatures. All have deep eyes and their skin is so thin they don't store or ship well, so the agribusinessmen don't like them either.

When we first tried growing potatoes some years back we bought **White Cobbler** "sets," eyes with thumbnail-sized chunks of potato cored out, dried and sold by some of the mail order seed sellers. If kept in the refrigerator till planting time, they stayed fresh enough to germinate at about a 50 percent rate. Production per set was nowhere near what we get with cut potato sections, but at least we tasted real potatoes for the first time in years.

SEED POTATOES

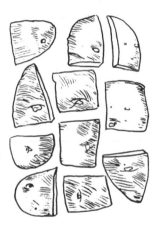

Cut up seed potato

And there's the cost. For what 300 sets cost we now get a hundred-pound bag containing about two hundred seed potatoes. It's certified, meaning it is guaranteed poison-free. So we plant a third of them and eat the rest. They are welcome too, as the end of our last year's crop always goes sweet around the middle of spring.

First I cull out the very smallest tubers, which will be planted whole. Larger potatoes are cut in chunks, each with at least two good eyes. All are dusted with natural sulphur and shaken in a paper bag much as you bread chicken. This is called "suberizing." It helps dry the cut ends, seals any nicks that might admit rot, and kills any lingering bug eggs. Some may object to this practice on the grounds that sulphur is a chemical. If so, it's one that occurs naturally in the ground and it is a vital element in a good many forms of life. I put it in about the same league as table salt. Too much of that will harm you too.

The suberized seed, about thirty-five pounds of it, is left in a warm place till the eyes show little nubbins of beginning sprouts. Then I put the furrowing attachment on the big tiller and cut six-inch deep furrows three feet apart along 150 feet of row. The potato seed is snugged down on the soil, the pieces a foot and a half apart, the young sprouts pointing upward. I scatter on just enough soil to cover the seed. As we discovered once too often, during any but an abnormally dry spring a thicker soil layer or a thick mulch will rot much of the seed.

The plants spring from the eye—the roots going down,

the shoots up. You can even plant deep slices of skin and eat all the potato, or so I hear. We've never tried it. As our sprouts grow and the clusters of tightly packed, deep green leaves poke above the ground, we scatter on old hay or chopped leaves. As the sprouts grow, we keep adding mulch to make the plants "reach". This way the tubers will develop just at soil level, or perhaps with their bottom third dug in, so that harvesting is an easy picking-up job, not the back-breaking grubbing needed when potatoes are grown without an easily removable mulch cover.

PROBLEMS   Using certified seed, rotating annually (but not following tomatoes so as to avoid wilts), and planting to an unlimed plot (to

The smallest child plays in the leaves as Daddy "plants" the early potato sets under an organic mulch that will make harvesting so easy.

avoid scab), we have few potato problems. Bug eggs or diseases that winter over go to food or compost; the vines are pulled and fed to the livestock or tilled under well. Cull potatoes are left in a pile to soften and are fed to the hogs. The one problem we do have are Colorado potato beetles, orange and black beetles with voracious, squashy larvae that are orange with little black spots on their sides. With earlier small plots we used to hand pick larvae and squash egg clusters. Now I resort to Rotenone, a short-lived naturally-occurring insect killer that is as close to organic as you can get. As soon as the larvae get numerous enough to be worrisome, I give the plants a weekly dusting, not spreading it everywhere but just where I find a gang of bugs. I suppose we have potato vine-borers and other pests, but they have never gotten so numerous that we noticed them.

Potato beetle

Broccoli

As April draws to a close I make our early planting of the cabbage family, our favorite member of which is broccoli. An LR Square is planted to the fairly recent introduction from Japan, **Green Comet**. February-started plants are moved out into the cold from the hotbed in late April and spaced a bit less than two feet apart so as to have four rows each way and sixteen plants. If the temperature is much below the hotbed's constant fifty degrees, I cover the plants with paper Hot-tent miniature greenhouses for a few days so they can harden off (which means they get used to fluctuating weather) slowly.

When they first introduced **Green Comet**, the **VARIETIES** seedsmen touted it as maturing only thirty days from planting. That's why we bought it the first year, and we were surprised and disappointed when it took exactly twice the advertised time. So were the seed sellers, and embarrassed too, I'll bet. They've had to up their estimates. Still, the Japanese hybrid is the earliest broccoli around and is a fine one. By the end of June most years the plants will begin turning out heads that often reach nine inches across, each main head providing a jumbo-sized family meal serving of tight-packed blue-green flower buds of superb flavor and texture. Then, starting as early as a week after the main head is cut, dime- to silver-dollar-sized secondary buds pop from the axials of every leaf up and down the stalks. So long as these buds are kept picked the plants continue to earn their compost, the plot producing at least one large meal serving, often two, per week till hard frost.

For the main, freezing crop, we use the old variety Waltham 29. Soon after the early transplants are in I pulverize about twenty square feet of soil in a location to the sun side of that year's pea patch. I put in about twenty-four feet of row, the birdshot-sized seed in quarter-inch-deep furrows six inches apart, with fine soil firmed well over them and plenty of water sprinkled on. The plants will be gradually thinned to stand two to the foot, so we end up with about four dozen plants to be transplanted into the pea plot once that crop is in.

Thus we get fresh, early broccoli from July to mid or late October. The late crop produces enough in September to provide at least two servings a week from the freezer for the balance of the year. We are always sure to let some of the early heads go slightly to flower, so a few bright yellow blossoms pop out from the green bud clusters. Eaten fresh, partially flowered broccoli has a fragrance and taste that is doubly delicious. The special quality doesn't survive freezing, however.

Fresh broccoli.

SAVE THE STALKS

When the plants are all pulled in October, most to be stored in the cold cellar for winter stock feed, we always save a few for several meals of broccoli stalk. The tough, woody outer layer of the upper two thirds of stalk is pared off and the light-green core can be eaten raw as a delightfully crisp and pungent relish. Sliced and steamed it is a crisp, Oriental-type vegetable, and it can be substituted for water chestnuts in Chinese dishes. It is best when it is left whole, parboiled for a few minutes in salted water, then covered with meatless spaghetti sauce, then with grated, aged Parmesan cheese and broiled till the topping is brown for *Broccoli Stalks Italien Louise*, named after its inventress. It's my Dad's favorite meal—nearly turned him into a vegetarian when Louise served it with our own rose' wine, a green salad, and homemade kneaded egg noodles. The latter are made by blending two eggs, a tablespoon of water or a bit more, with enough flour to make a thick dough. The dough is kneaded for a quarter hour or till the gluten is worked out of the flour as with bread dough. It's then rolled, sliced to noodle proportions, and dried.

Cabbage worm

Broccoli is an undemanding crop, requiring only good soil and a mulch which is put on when the tomatoes are planted and mulched. Liming and rotation keep clubfoot away and one of our biological controls, the trichogramma wasp parasite, is nearly 100 percent effective against cabbage loopers. The trichogramma are used primarily for the orchard and are discussed in some detail in Chapter Six.

Cabbage

We aren't great cabbage fans and since this plant needs a lot of special attention we don't plant a whole lot. I usually have a dozen or so seedlings of **Jersey Wakefield** early cabbage in an odd corner of the garden. This variety matures in about two months and as cabbage goes it is sweet and mild in slaw or steamed for a hot course.

When I put in the nursery rows of late broccoli I usually plant a row or two of the late variety **Chieftain Savoy**. This crinkle-headed cabbage takes a good four months to mature on our land, but is mild enough eaten fresh, will store through the winter, and makes good kraut.

Cabbage, like broccoli, has a shallow root system that really appreciates an old hay mulch from transplanting time on. Unlike broccoli, it needs a lot of protection from bugs. Just one looper or imported cabbage worm that evades the trichogramma can ruin a young cabbage head. Even worse are the striped flea beetles. They come on in hordes and can make a cabbage leaf look as though a load of buckshot has passed through. Supposedly rotenone will control these tiny pests, but I don't like to put poison of any kind directly on a part of any vegetable we are going to eat. It stands to reason that cabbage, which grows by sort of wrapping itself up in its own leaves, would wrap up some of the poison too, doesn't it? Rotenone may lose toxicity rapidly and the makers claim it is harmless to humans. But it kills bugs, and it kills fish, and it kills hogs, so I don't see how it could be good for our kids. Using it on potato vines is one thing; cabbage is another thing altogether.

PEST CONTROL

So, I plant cabbage in double rows, making a four-foot-wide, long and skinny cabbage patch. As soon as the plants' outer leaves begin to show bug damage, so long as they've not begun to head, I put on a light dusting of rotenone—these leaves wouldn't be eaten anyway. Then I surround the row with old hay bales and lay over them a length of transparent plastic window screening, held down with rocks. I suppose the screen cuts down on received sunlight and slows growth somewhat. But

DOUBLE ROW PLANTING

it cuts down on the bugs too. Any that slip through are handled by the toads we invite to take up housekeeping under the screen.

Cauliflower

We don't have very good luck with cauliflower. The white varieties apparently need cool weather along with strong sun to head properly. In our dry, hot summers they put out little button-like heads before they are fully mature. Planted to mature in the cool fall, they don't head at all.

We do get fair results with **Purple Head** cauliflower, however. We use the latest-developing variety available and plant it just as the late broccoli. Heavily mulched, given water in dry spells, and with luck, plants will put out six-inch heads that turn bright green in cooking, are good raw as a relish (once you get over the purple color) or in salads. The flavor is blander and the texture finer than either broccoli or white cauliflower and it's so good we keep growing a dozen or so plants every year, though half the time they fail to produce satisfactorily. Maybe with more experience we'll hit on a few tricks to grow bumper crops.

Kale

Kale isn't worth eating till it has been tenderized and sweetened by a few October frosts. But then it's just about the best green around. It's a slow grower, though, so I put in a complete LR Square of transplants from the nursery frame and seed in rows a bit more than a foot apart, plants a bit less than a foot in the row. The hotbed-started transplants are more for goat bait than anything else. When these potential garden-wreckers get loose and head up the garden path we know the kale will hold their attention long enough for us to come to the rescue. The kale doesn't mind—it just puts out a new crop of leaves.

Kale also gets planted in any vacant garden space through the month of June. It grows slowly, but by fall frost we will have perhaps a hundred kale plants, each with a full complement of crisp, crumpled leaves. In late fall, most of the plants are pulled to join the old broccoli plants and cabbage in the cold cellar. Each winter day the goats and rabbits get a ration of fresh greens to supplement the dry feed.

COLD STORAGE

Kale is probably the most cold-resistant plant grown. I always leave a bloc of plants in the ground (some years with parsnips along side) and in late fall, the same time I'm stacking insulating hay around the bee hives, I surround the kale with bales. Boards go over them and so long as someone is willing to dig down through the snow, we have garden fresh greens on the table. Most years the kale is eaten by January, but more than once we've had plants live through the winter and begin growing all over again in spring.

Carrots

All the books tell you to make repeated sowings of carrots till midsummer, since left in the soil too long the roots are supposed to get woody. Ours don't. The experts also say to delay planting till after June 1 to avoid the carrot root fly, a long-legged relative of the house fly whose maggots can dig into a carrot, admitting a rot that ruins it. We don't have that problem either.

In late April, as late as possible, I plant two LR Squares of carrots. Each fall I till a garden cart of sand into the future carrot plots to help the roots grow long and straight. Still, our soil is shallow enough that we'll never do well with those foot-long carrots they grow in California. We stick to the shortest varieties available—one square each to **Nantes** and a **Chantenay Variety**. In shallow furrows a foot apart I sprinkle seed *very* thinly, add in two or three radish seeds per foot and then use the planting board to tamp the seed into the soil. Then, rubbing soil between my palms, I just cover the seed and pat it down gently. Planted this way and kept moist, the radish seed sprouts, marks the rows, and breaks the surface for the tiny and slow-germinating carrot seed.

The carrot squares get a weekly weeding and thinning till the radishes are big enough to be pulled and eaten, about the time biting bugs are getting unbearable. After a final weeding and thinning, so plants are separated by two inches, I put a thick hay mulch between the rows, but leaving the carrot crowns exposed for treatment against the root fly. There is a spring hatch which never does any appreciable damage—if they stunt or kill any of the tiny seedlings they are just helping with the thinning. The fall hatch is something else, though. All during August and early September we keep a windrow of loose wood ash heaped over the crown (but not enough to cover any leaves) of the carrots. New ash is applied after the infrequent, heavy rains and it works. No

ROOT FLIES

one knows just why the ashes keep the female flies from laying their eggs, but it does. And the ash contributes its potash to the soil with the fall tilling.

We pull carrots as we need them, beginning with the tiny ones that are fleshed out enough to eat by early June. The inter-row mulch keeps the soil cool so growth is slow but constant, and when the roots are dug in late September, they are just as sweet as they were in June. And that's a lot sweeter than any carrot grown commercially. You don't have to peel them either, just scrub them hard with a vegetable brush. Of course, a lot of our carrots have to grow around stones so not a few are mighty peculiar looking. But they are delicious, and stored in moist sand in the cold cellar they stay that way till spring.

Onions

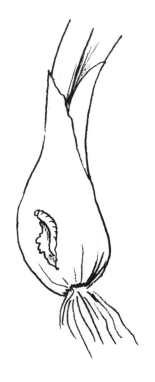

Onion root maggot

The last spring planting of any consequence is onions. The seed requires over 100 days of prime growing season which we don't have, so we plant sets. **Stuttgarter** is a new and I think imported set that is smaller but much better than the old **Ebenezer** sets that tended to either sprout or rot in storage. I get five pounds of these sets from the local farm co-op and plant two LR Squares. The largest sets are punched down into shallow furrows, sets three inches apart, rows six inches apart. Soft or dried out sets are discarded, the very small ones set in between the larger ones to be pulled for spring green onions. In early spring the onions get the same weeding and mulch treatment as the carrots. But once they are well up I forget them, only pulling perennial grasses and large weeds that might shade the onion plants. The bulbs tend to the small side with such weed competition, which is fine since they seem to keep better than larger onions.

I've heard that there is an onion root maggot that we would control with wood ash as with the carrot pest, but we've never seen evidence of it or any other onion problem. Either the bugs don't come this far north or all the weeds growing in the onion bed camouflage their favorite garden prey. Pretty ''iffy,'' that theory, but there may be some sense to it.

It is never too long after the onions are in that several small black buzzing dots appear to begin dancing in front of my eyes—the first black flies announcing that biting insects are taking over the outdoors for the next few weeks. So I hustle through a final weeding, apply the mulches, and abandon the outside to the bugs. There is plenty to do indoors, working on equipment in the screened workrooms or with the animals in the barn behind a protective screen of spider webs.

Scattering potash-rich wood ashes along the well-mulched bean rows to improve yields. Ashes on crowns of root crops also prevent damage from several kinds of root maggots.

5

Small Stock
the Easy Way

With the arrival of May the days are getting long enough that our pasture grass is beginning to grow, so it won't be too long till nature will take over feeding an expanded summer population of small livestock. Over the coming months the land will permit us to stock the freezers with a twice-weekly feast of plump, tasty chicken, as many fryer rabbits, and an occasional domestic duckling or pheasant.

Pets and Pests

You may think it odd that I am starting this chapter with a discussion of pest and pet animals. But as we discovered when the German Shepherd "liberated" our first trio of breeding rabbits from a too-weak cage and a nesting goshawk selected the chicken pen as its meat market, all of nature's creatures must be considered before a new homesteader goes into livestock.

In the old days most farmers kept both dogs and cats, the former to chase off chicken thieves, the latter to catch rodents. Today's environment-conscious homesteader may go along with the canines, but cats are something else. It's a too little-known fact that the animal most responsible for destruction of wild birds and small animals, beside man himself, is the common house cat. Certainly no one can complain about Aunt Agatha's house tabby that stays inside all her life. But even the lovely Angora or Siamese that purrs and rubs against your leg to be let out in the evening becomes a ruthless killer on its nocturnal rounds. Even more destructive are the feral—gone wild—cats, the result of thoughtless people who let their female cats breed at will.

There are several new city-based organizations dedicated to neutering pet animals, which is a good step toward stopping the urban and suburban cat population explosion. In the country, though, wildlife population is benefited in direct proportion to reduction in the cat population. We do our part by not keeping cats. And, after witnessing a stray feline destroy a brood of tiny ruffed grouse chicks one spring as the mother squawked helplessly nearby, I have declared open season on wild house cats.

CATS AND RATS

There's no question that cats will keep rodent populations down naturally, and without them you have to take special measures to compensate. I'm reminded of the incident, in Ceylon I think it was, just after World War II when the local government dosed a large area with DDT against mosquitoes. The lizards had had a feast on the dead bugs and weren't hurt, but when the cats ate the lizards the accumulation of poison was too much and the cats all died. Then the rat population exploded—poison couldn't keep up with the rat birth rate—and finally the Air Force had to parachute in several plane loads of house cats. Needless to say, the DDT program was cancelled too.

We avoid a rodent population explosion in other ways. Basically, we do everything reasonable to deny rats and mice a food supply. All feeds are stored in metal cans or creosoted wood bins that the rodents won't chew through. Holes that are occasionally gnawed through barn walls into pens are stuffed with steel wool, or if large enough are closed with a square of aluminum roof flashing nailed on well. Feed troughs are never filled so full the animals will spill, and we are careful to drop as little feed ourselves as possible. What does get scattered around is eaten by the bantam chickens, and in the winter the barn is frequented by blue jays after any stray grains.

We have never seen any Norway rats, the immigrant that infests the cities. I assume they would have too long a hike to get to our place. The occasional wood rat, a native forest dweller, and the plentiful white-footed and meadow mice that try getting into the house and barn each fall are taken care of with a few spring traps. In rodent control as in everything else we try not to upset the natural balance, for example by not storing corn in traditional slat-sided corn cribs which would provide the unnaturally large feed supply that permits an unnaturally large rodent population. With just a little added effort we can do it without the help of cats.

Felines No, Canines Yes

We have about as many reasons for keeping dogs as for not keeping cats. For one thing, any homestead in an isolated setting is easy prey for the human pests that seem to be on the increase in these troubled times, even in rural areas. Really big dogs with a good voice and all their own teeth are about the best deterrent to people problems I know of. Our German Shepherds submit calmly to the worst ear-wrenchings our three-year-old son can dish out, but to uninvited visitors they are bristling, fanged terrors. None of our dogs has ever bitten anyone. We don't train them to *be* vicious, just to look that way, but then it's a rare stranger who is foolish enough to approach the home place without the dog's and our invitation.

DOGS AND GARDEN PESTS The other reason for keeping dogs is protection of another sort—from garden marauders. Rabbits, groundhogs, and deer can ruin a young garden and opossums and raccoons love to harvest the sweet corn just a day or two before it reaches its prime. Shortly after we moved onto the place I took the dogs on a leashed walk around the orchard and gardens-to-be. After several such tours they accepted the area as their home turf, to be continually prowled over and protected from interlopers of any sort. The fresh dog smell alone keeps the deer well back into the woods. The occasional groundhog or rabbit that ventures beyond the stone walls into the lettuce patch is usually woofed away before any serious damage is done, and a good many of them inadvertently help with our budget by reducing the dog food bill for a day or so. In fall if there is evidence of possum or coon activity in the corn, I shut the dogs in for a night, set traps, and augment the budget even more directly with a fine meal for all of us and perhaps a new coonskin Dan'l Boone hat for the youngest hunter in the family.

Another kind of left-handed advantage of having big dogs on a homestead is that, unless you are a lot better dog trainer than I am, all pens and cages have to be big dog-proof. But any fence that will keep the Shepherds out will also restrain the foxes, weasles, mink, coyotes, and stray dog packs that wander in our woods and would happily dine on prize laying hens if they could get at them.

The Chicken Flock

Of the animal foods we raise, the difference between home-grown and store-bought is greatest in the eggs and poultry. Most city people don't realize it any more, but a fresh-laid egg has a yolk that is deep yellow, almost a pinkish orange. The white holds together, the whole thing sits up fat and sassy in the pan, and it tastes like an egg. By comparison supermarket eggs taste flat if you are lucky, almost spoiled if you aren't.

STORE EGGS The problem is age; most store eggs spend weeks or months in cold storage. In commercial "egg factories," even though they live in cages and never lay eyes on a rooster, the hens think their eggs are for brooding into chicks. Or they do with what is left of their natural instincts; the hybridizers have just about bred the maternal "broody" instinct out of the chicken. In any event, like most of nature's creatures, hens lay most eggs in the spring, so as to have young fully grown by cold weather. The resulting spring glut of eggs is stored and parceled out to stores over the year. During up to six months of cold storage these infertile eggs are doing the same thing they'd do if left in the hot sun, only they are doing it more slowly. Natural decay enzymes are breaking down the once-living tissues and water is slowly evaporating out through the shells that have been scrubbed "clean" of the natural preservative coating the hen gives it. About the only good thing about stored eggs is that they will peel easily when hard-boiled. Our fresh eggs haven't lost any fluid so they are unshrunk and pressing so hard against the shell that peeling a hard-boiled egg removes a layer of white. Before making egg salad Louise has to let eggs sit around and age a few days so they'll peel.

STORE CHICKENS If commercial eggs are bad, commercial poultry is worse. The breeders have succeeded in developing strains of birds that grow up in only six weeks of living on chemicals during which time they never have a chance to be chickens. They live out their brief span in a totally controlled artificial environment, never breathe fresh air or see the sky or feel the warmth and health-giving rays of the sun. And to top it all off, they are sent off to market before they even know whether they are hens or roosters. Small wonder they don't taste like anything.

In contrast, our chickens live a full chicken life, a fact that isn't always appreciated when the crowing contest begins around 3:00 a.m. on a summer's morning. But they *taste* like chicken should, with a full, rich buttery flavor that no amount of "flavor enhancers", MSG, or bread crumbs laced with some secret blend of herbs and spices can impart to a commercial bird.

Choosing a Chicken

FOR EGGS

There are five basic poultry types to choose from, and in our decade or so of experimenting we've tried most of them. *Egg Producers* such as the White Leghorn (pronounced "leggern," by the way) and hybrids based on that breed are white-feathered, skinny and temperamental little birds. In the egg factories they lay some 200 large, white-shelled eggs in their sixth through nineteenth months of life, after which they give up and get sold for cat food. They have meat, of course, but not much, and what there is is pretty stringy.

FOR MEAT

Meat Producers are bred to flesh out rapidly with maximum feed economy. Most these days are hybrids, the best known being a cross between the White Cornish and White Rock breeds. Slaughtered at only four weeks of age, they are sold as Rock Cornish game hens in fancy restaurants.

White Leghorn

Bantam

Dual Purpose hybrids of several kinds are common in New England where they produce jumbo-sized eggs with the brown shells that Yankees prefer for no particular reason, and they develop into fine eating at roaster size. Their main disadvantage for a homesteader is that they don't come true from the egg, though in our experience, the hens of these hybrids show more maternal instinct than any other full-sized breeds. However, if the hybrids' eggs are hatched, the offspring will be all colors of the rainbow and if let breed themselves would probably exhibit atavism—reverting in a generation or two to the scrawny, unproductive multicolored creatures that were the original wild chicken stock.

FOR MEAT AND EGGS

A fourth category of chicken includes a variety of fancy breeds kept for show or hobby. Some have odd coloring or plumage; there's one with a bare neck like a turkey and another South American breed called Araucana that lays olive-drab colored eggs (the fancy fowl merchants promote this breed as "layers of pastel-colored Easter Eggs"). The fancy breeds aren't worth their feed bill so far as meat and egg production goes—they are hobby animals, and I can't rightly call cleaning out a chicken house entertainment.

FOR FANCY

The only novelties we keep are bantams—pint-sized

Barred Rock

Rhode Island Red

versions of standard breeds. Bantams seem to have more sense than larger chickens and they are a lot better at flying so we let our flock run more or less free. They do a good job of scavenging up spilled grains, digging fly larvae out of odd barnyard corners, and keeping down the bug populations in the gardens. It's fun to watch them roost in the trees in the evenings.

OUR CHOICE   The main flock consists of the final variety of chicken, *Purebred Multipurpose*, according to the hatcheries. These include the well-known heavy-bodied breeds, Rhode Island and New Hampshire Reds, and the White, Plymouth, and Barred Rocks. Any of these breeds will produce good brown eggs and fine meat, but not as efficiently as the hybrids. Most important to us, they come true from the egg, which means we can have a self-perpetuating flock, or could if the commercial hatcheries hadn't bred the nesting instinct out of the strains. (Later we'll get into our ongoing attempt to breed the brooding instinct back into the birds.)

For some years we kept Rhode Island Reds, a hardy and productive breed with gorgeous mahogany-red plumage. However, the flock developed a congenital club-foot-type deformity. Next we shifted to a hybrid multipurpose breed that was a tremendous egg machine but couldn't reproduce itself. Presently we are running mainly Barred Rocks. Their feathers are black and white, temperament calm, meat and eggs good.

Equipment

Plenty of homesteaders let their entire poultry flock run loose to forage for much of their food, at least during warm weather. If you don't mind chicken droppings all over and having to gather eggs from some of the odd places that hens pick to nest (the air cleaner of your truck or inaccessibly far back in the wood pile), that's all right. It's far better, though, to build a closed pen and henhouse.

OUR HENHOUSE   Our winter flock of a rooster or two and a half-dozen hens, the bantams when they are of a mind to come in out of the cold, as well as the summer flock of thirty or more birds, is housed in a henhouse only three by nine feet in floor area with a ten-foot-high ceiling and an outside run that is nine by twelve feet in area. This layout may appear small. However, the total area of about 130 square feet could theoretically carry up to 400 chickens if they were caged as in many egg or broiler factories. In factories where the hens are let run loose, they still provide no more than one to one and a half square feet per animal. So I figure our chickens have it comparatively roomy. About the only problem comes late in the summer when the expanded flock is nearing slaughter weight. There can be quite a squawk evenings as all those birds bustle around after a roosting spot in the henhouse. When the squabbling lasts more than a minute or two we know it's time to get out the scalding pot.

A homesteader can make a henhouse from just about any outbuilding that has a varmint-proof floor, sound walls and roof. In the North it should be relatively free of winter drafts; in the South the main requirement is summer ventilation. We put our henhouse in a long, narrow, and unused corner of the story and a half shed that juts out from the southwest corner of the barn. I cut a yard square hole in the outside wall about five feet off the ground and stapled stout aluminum flyscreen over it. The opening lets air circulate freely, admits some of the setting sun, but keeps flies out and chickens in. **DESIGN GUIDELINES**

The birds prefer to spend the night as far off the ground as they can fly, which isn't very far. Rough lengths of two-by-four-inch lumber are nailed at the five-foot level along each of the long walls. On these, extending across the short (three-foot) dimension of the house are nesting boards—one-by-two-inch rough lumber, the back roost six inches out from the back wall, the others a foot apart. In summer there are a half-dozen roosts, in winter only three or four for the dozen or so birds. The books say a heavy-bodied chicken requires a good square foot of roosting space, but I've seen up to a half-dozen birds squeeze onto one of those three-foot wide perches and appreciate the togetherness on a cold night. **ROOSTS**

Roosts must be rough and fairly wide so chickens' big, awkward feet can get a good grip. A round, smooth dowel like an oversized canary perch wouldn't work, though a length of rough-barked tree limb would be fine so long as it was at least three inches in diameter and fixed well so it didn't roll when the birds landed. Roosts are needed only for mature birds, by the way. Youngsters can't fly during their first three months or so. Most years, by the time our spring hatch has fledged out well enough to get up to the roosts they are comfortably in the freezer.

At floor level under the window is a hole a foot square with a door that slides up and down in rough wood channels like **THE DOOR** the blade of a guillotine. It operates with a length of woven steel wire that extends from a bent nail in the top of the door, up through a hole in the window sill, and out by the doors of the henhouse. The door, opening out to the run, stays open most of the time but it is handy to be able to close it when it's time to keep the birds out to clean the house or keep them in to be caught for dinner.

The roost in our henhouse is this simple.

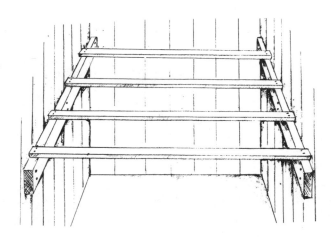

The ceiling of the henhouse was originally open to the shed loft. To keep the birds in I laid several old window shutters across the ceiling beams. Their louvered openings keep the birds in but allow hot air to rise, carrying with it a continual cargo of "chicken dust" which will mean a big cleaning job when I find time to turn the upper story of the loft into a pigeon loft. A pair of stout shutters also serves as doors to the henhouse, hung batwing style on bent nails. No latch I was able to fashion was stout enough to keep curious dogs out but let curious kids in, so now the doors are kept shut by a length of two-by-four-inch scrap lumber laid diagonally across with the ends wedged between upright studs on either wall.

Henhouse Furniture

Besides eggs and drumsticks, chickens provide copious amounts of another valuable product—manure. It's wonderful stuff once tilled into the garden. But when it collects in nest boxes, ruins feed or water, or sticks to the soles of an egg-collecting Grandma's bedroom slippers, it is horrid. After years of experimenting we've come up with a layout of henhouse furniture that pretty well keeps the droppings where they belong. The nest box is attached to the side walls at the top of the five-foot-high doorway; the doors are its rear panel, and to collect eggs, we just open the doors and reach in. It is a simple rectangular box three feet wide by eighteen inches deep. The front panel is a foot high with a round hole ten inches in diameter cut in it. The roof slopes up and back from the front panel at about a forty-five degree angle so the birds can't roost on it. A roosting board is attached to the front, extending six inches out from the front so the hens can get in easily. This single "dark nest" is kept full of fresh straw that we fluff up each time eggs are collected. The hens like to lay eggs in the dark corners, sort of burrowing into the straw; I suppose they think it is a safer place than an ordinary open-front nest. It will handle up to four hens at a time, each snuggled back into her own private burrow.

Suspended from the nest bottom is the self-feeder, the sort that is available from any feed store or farm equipment dealer. It is nothing but a galvanized sheet metal cylinder, ten inches in diameter and two feet high, with a shallow pan hung from its bottom. The pan is several inches off the floor so it will swing away if a bird tries to scratch feed out, thus wasting it. However, they can peck out feed at will. The feeder holds up to fifty pounds of dry feed, enough to last the winter flock for weeks at a filling.

Water takes a little more effort. Fresh, clean, and unfrozen water must be available to the flock at all times. Probably beginners lose more birds from a lack of water than from any other single cause. And just a few hours without adequate water can lower egg production by as much as twenty percent. We use a five-gallon-capacity dome-type waterer. This is made as if you took the feeder and welded the pan and cylinder together and punched a hole through the cylinder at its bottom. There is a spring-activated plug that closes the hole when the dome-topped outer cylinder is removed. This top is another cylinder that is closed at one end. You pour water in the base, and put the outer dome on to compress the spring-plug. Water trickles out into the pan till it covers the hole. A vacuum is formed inside, and the only time more water will flow out is when the chickens drink enough that the hole is uncovered so that a few bubbles of air can get in and a few dribbles of water flow out.

THE NEST BOX

Self-feeder

FEED

WATER

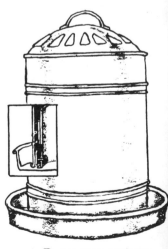

Dome-type waterer

In summer I keep a hose at the back of the barn, making it easy to clean and refill the waterer as often as needed. In winter, freezing is a problem and since water must be carried from the house, I don't like to see it wasted. We have found that the best way to keep the waterer clean is to put it on a foot-high platform under the nest box and beside the feeder. The birds have to stretch to get a drink, but they can't scratch litter or get droppings in the water. The stand is a rough wood box; on it is the heater which supports the waterer. The heater is an over-sized metal pie plate with a low-power electric heating coil in it. Plugged in all winter, it keeps the water flowing on all but the coldest subzero days.

Heater for waterer

The Outside Run

A chicken benefits from healthy outdoor exercise as much as any other living creature—maybe more so. The sun provides them with vitamins A & D just like you or me. And besides, they get grit for their gizzards and trace minerals for general health by pecking around in the soil. Dust baths provide a fine, natural defense against lice and other external parasites, and bugs, leaves, and grass provide a healthy supplement to their usual diet. Besides, chickens were created to be active, outdoors critters. Caging or cooping them permanently isn't natural, and when they are treated that way they act peculiar, most commonly pecking one another to the point of cannibalism.

THE RUN

Our birds' outside run is contained by standard poultry netting strung from the shed walls out around two stout posts sunk a good foot and a half into the ground. As with all fencing, I dug the post holes two feet deep and a foot or more across. On the bottom I put a flat rock, the post went on top of that, then the rest of the hole was filled with good sized rocks, wedged in tight. I mixed a slurry of Portland cement, sand, and enough water to make it runny enough to flow well. Poured into the hole, the slurry binds the rocks and posts into a solid anchor. The wood is fully sealed from water by concrete and stone so that it will last, especially if the sub-ground-level portion has been soaked in a bucket of creosote for a week or more. Treated this way, most any hard wood will do for posts, though red cedar, cypress, and redwood all have natural anti-rot protection built in.

Chicken netting is pretty floppy stuff and can't be

First step in taking on animals was to build the chicken pen at the back of the barn.

stretched tight like wire fencing so it must be framed in. The top of the five-foot-high net is secured to stout poles running from the corner post to the shed walls. Along the bottom of the run I buried six-inch boards; the bottom of the net is stapled to them. Besides anchoring the net, the bottom boards discourage any dog or wild critter that tries to dig in for a free meal.

Our heavy-bodied standard-sized birds can't fly very well, but they can get to the top of the run sides. They used to be content just to walk around on the poles, but when we took on bantams that fly all over the place we got problems. The big birds would jump off the fence trying to imitate their smaller kin, in the process landing with a thud that broke a leg or getting so winded they were too easy prey for the dogs. So, I put chicken wire over the top of the run, just laying it on but leaving a small space at the end open enough for the bantams to get in and out.

Bringing Up Chicks

Before the breeders got to them, all hens went broody every spring. They'd begin fussing and clucking more than usual, refusing to leave the nests, and generally getting maternal. Left to their own devices they would lay a clutch of about a dozen eggs over about a two-week period, then settle down to brood them so they hatched at the same time. Once the chicks were grown, the

At its beginning, a good chicken dinner won't fill the hand.

hens would moult—change feathers for the coming winter—and wouldn't begin laying again till the following spring. Over hundreds of years humans have been breeding this natural cycle out of chickens by selecting those hens that showed the least tendency to brood and the greatest inclination to lay over longer and longer periods. What we have now are hens that begin laying in their fifth to seventh month of age, skip the broody period and the moult, and lay more or less continually for a year. In commercial operations they are sold as fowl when about a year and a half old—when they are still laying but are about to become less productive.

So our twentieth century chickens have forgotten how to reproduce themselves. Their continued survival is dependent on hatcheries that take fertile eggs from the best producers, brood them in incubators, and sell day-old chicks to the egg or broiler factories. Louise and I think this is just one more example of technology interfering with nature for the sake of efficiency and to make someone a few extra dollars they probably don't need. In the process not only the egg and poultry consumers but the entire chicken race has become dependent on that technology and its possessors. Mainly out of pure ornery independence, I guess, we are doing our best to breed the old survival instincts back into our birds even if we may not get quite as many eggs per chicken as someone else. More on that later.

The basis of our whole chicken program is a dandy little gadget called the Marsh Turn-x incubator form Marsh Farms, Garden Grove, California 92240. Ours cost about twelve dollars some years back. It holds about a dozen eggs, and keeps them warm, properly humidified, and turning, the same as if they were under a hen. Several times a year, depending on how much chicken we've been eating, I'll collect the dozen biggest eggs laid over a two or three day period and brood them for the three weeks they need to develop. Usually ten will hatch, sometimes only half the dozen. But these chicks will stay warm in the incubator for up to a week till I get the regular brooder fired up.

BROODING EGGS

I suppose that any heat source that will keep a two or three square foot area at about 100°F. would do to brood chicks. We started out with nothing but a big 200-watt bulb in a cone reflector made of aluminum foil. Now we use a commercial electric brooder, an aluminum hood the size of a large mixing bowl holding a heat-proof ceramic socket that uses a standard heat lamp. (The red glass, shatterproof, infrared kind of lamp, not a "sunlamp" that generates ultraviolet suntanning rays that would literally sunburn the chicks alive.) We got this outfit from a mail order house, but any feed store would sell them for five dollars or so.

During their first week or so, chicks don't stray far from the heat which imitates the mother hen's body warmth, and they

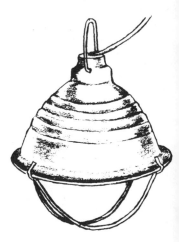

Electric brooder

A dozen eggs in the Marsh incubator, where they spend three weeks in proper heat and humidity, needing only to be turned three times a day. Anywhere from the nineteenth to the twenty-second day an egg begins to peep, then the first pipping chick breaks through its shell. After several hours—sometimes as much as a day or two—of pipping, the chick will kick itself free of the shell and in another couple of hours will dry off to become the cute ball of fluff that is a day-old chick.

can just about be brooded in a shoe box. Indeed, the first chicks Louise and I raised up spent their first few days on the kitchen table in an old orange crate. They provided the evening's entertainment till we discovered that even tiny chicks raise a continual cloud of dust which in time coats everything with a fine layer of whitish, highly caustic powder. So the birds went outside and, since it had been my idea, I couldn't offer much of an argument when Louise suggested sweetly that spring cleaning should be a family affair. It still is.

A cardboard box, egg-carton feeder and brooder bulb will serve to house the new chicks for the first week or so.

Now, we brood in a box eighteen inches wide and four feet long that can hold up to thirty chicks. It's a foot high and has a length of chicken wire stapled along one side. The wire stretches tight over the top and cinches onto several nails hammered into the far side to keep pests out and all but the most adventurous chicks in. I doubt that the chicks are in much danger from wild critters so long as the brooder light is on, and strange as it may seem, house cats will never bother chickens. More than once, though, I've seen a big dog peering curiously up at the table which supports the brooder, so the wire is mainly a dog protection.

OUR BROODER BOX

The heater light must be suspended (we use wire wrapped around a nail in an overhead beam) so the temperature at ground level at first is just about 100°. You can use a thermometer, I suppose, but the chicks are the best indicator of whether the heat is right. If it's OK—during the first week our lamp bottom was just eighteen inches from the brooder bottom—the chicks will cluster around the edge of the reflector. If it were too cool, they would gather right under the bulb; if too hot, they would keep away. As they grow insulating fuzz, then feathers, they need less heat, and I raise the brooder exactly two inches each week, removing it completely at the end of their fourth week unless it's winter. During cool weather, I cover each end of the box with sheets of cardboard and leave the brooder light on, the bottom of the bulb two and a half feet from the box bottom. On the rare occasions we brood in the depths of winter I do the same, and also lay a length of hotbed soil-heating cable in the litter in the box bottom to keep the chicks' feet warm.

Brooder Furnishings

The brooder floor must be covered with some sort of litter to absorb the semiliquid droppings. Any organic matter such as straw or crushed corn cobs would work, but I don't like anything flammable that close to a heat lamp in our old wood barn. So I use sand, a good two inches of it, in the bottom of the brooder box. It absorbs well enough, the birds can scratch happily in it and ingest it into their gizzards where tough muscles continually grind the grit and feed to aid digestion. The term "scarce as hen's teeth" means what it implies. Chickens require a constant supply of coarse sand or small stones to serve as "teeth" to grind any feed but commercial mash, which is pulverized by machine so the big

GRIT

egg and broiler growers won't have to bother supplying their birds with grit.

HOMEMADE EQUIPMENT

At first we fed and watered chicks with homemade equipment. The feeders were made from three-pound coffee cans with both ends out, a plastic ice cream tub and several lengths of coat hanger. One piece of coat hanger was bent double and pushed through holes in the top of the coffee can to serve as a handle. Two other wires were run through the can and tub in an "X" pattern, the can held inside the tub with its bottom rim about a half-inch above the tub bottom. When feed was poured into the can, it ran out into the tub as used and was an effective self-feeder.

Waterers were made of more ice cream tubs set upside down on shallow bowls. I cut a notch in the side of the tub, the notch a little shallower than the bowl was deep. The tubs were filled with water, the bowls placed on top. Then with the two held tightly together, they were flipped over with just a bit of a splash. Put in the brooder (close to the lamp in cold weather to prevent freezing), the water trickled out as the chicks used it, but was kept from flowing all over by the internal vacuum, the same as the dome waterer. It is a good idea to place a heavy object on top of the tub to keep the chicks from dislodging the water container in their continual running about. This waterer must be cleaned at least twice a day, as the birds get litter and droppings in the bowl within an hour or so. After the first batch of chicks, we broke down and bought a much larger, regular self-feeder and a dome-type waterer. Placed on an old pie plate, the waterer keeps the watering lip high enough so that the chicks can't track through it and cleaning becomes a less frequent chore.

A homemade feeder and waterer

More Observations on Starting Chicks

Always preheat the brooder, litter, and the chicks' water for several hours before they are transferred to the brooder. If you're not using sand for litter, supply grit of some sort in a separate dish.

FIRST FEED

For food, either use medicated commercial chick starter or add a mild intestinal medication to the water. We tried fine-grinding home-grown feeds and raising all-organic chicks and it just didn't work. There are a number of diseases that can wipe out a batch of chicks overnight—the intestinal ailment coccidiosis being the most common. If you miss noticing one or two chicks with loose, bloody droppings one afternoon, the next morning the lot can be dead. I can only assume that natural resistance to these diseases has been bred out of the chickens, so until we can breed it back in (if we can) I will feed young chicks standard medicated chick starter for their first month of life. A chick will eat about two pounds in that time, so a 100-pound sack costing ten or so dollars lasts us a good year or more. I figure that anything harmful in the bought feed gets worked out of the animal by the time it produces eggs or meat in another six or seven months.

FEED REQUIREMENTS

Probably the most important observation I can make for anyone beginning with poultry is this: in introducing stock of any sort onto our place we always try to find the healthiest—guaranteed healthiest—available. Trying to maintain a healthful environment with natural methods, it makes sense to keep out diseases that commercial farmers battle with an endless variety of chemicals. Then we can expose animals that are healthy at the start to the natural conditions that build natural immunity to natural problems.

With chickens, *don't you dare* start with anything but newly-hatched chicks (or hatching eggs) from a reputable hatchery whose flocks are registered and guaranteed *U.S.-Pullorum-Typhoid Clean*, in exactly those words. That way you are avoiding the two worst poultry diseases right off the bat. If there are other chicken flocks within a quarter-mile or so of your place, or if chickens have been run on your land within the past five years, it is a good idea to have the chicks vaccinated for other common diseases. The federal County Agent (if you can find him) or your state university's Agriculture Extension Service can tell you which diseases are prevalent. In our area most hatcheries offer vaccinations against Newcastle, bronchitis, and Marek's disease. The three add something less than a dime to the cost of each chick.

BUYING CHICKS

Commercial hatcheries seldom sell less than twenty-five chicks in a lot. Bought ''as hatched'' or ''straight run'' (50 percent of each sex), twenty-five will cost ten or twelve dollars post-paid to the nearest post office. A shipment of all pullets (young hens) of laying or dual-purpose breeds or all cockerels (young roosters) of broiler-type birds will be somewhat more. If the hatchery is out of mailing range, you may have to pick up your chicks at the nearest air freight office, which means a trip to the commercial airport nearest you.

However, no matter how much bother it may be to get

day-old chicks from a No. 1 hatchery, it's worth your time several times over. They will be completely problem-free—raised in a sanitary incubator, free of disease, parasites, or hang-ups of any kind. Now, poultry wholesalers or your neighbor might sell you a half dozen "started pullets"—young hens ready to lay. You are saving a few month's wait, but you may also be buying trouble. One of our own neighbors got "started" birds from a supposedly reputable source, and after just a few months all their feathers fell out and he had to destroy the whole flock.

Feeding and Management

CLEANUP

As each batch of chicks gets toward the end of its fourth week of brooding, I shoo the adult flock out into the run and close the little door into the henhouse. After removing all nests, feeding and watering equipment, I partake of a fiery potion especially prepared for the occasion and set to cleaning out the henhouse. The roosts are scraped, then I use a sharp spade to cut out the packed litter and droppings on the floor—up to a foot thick if it's gone for more than a few months without being cleaned. Though totally odorless at the start, once dug into, the litter gives off a subsurface perfume that will bowl you over. Primarily ammonia, but the pure organic kind, it's an effective remedy for a spring cold, though.

All scrapings are hauled to a garden or dumped on a compost pile. Then I hose the inside surfaces of the house well and spray with a strong solution of Chlorox and a strong industrial cleaner. All equipment gets similar treatment plus a scrubbing with a long-handled brush if needed. Once the water has drained off, I open up all the barn doors and windows to get the most drying effect for the wet wooden walls and floor and go off fishing or to the garden—anything that won't remind me of chickens.

A NEW HOME

Next day I put an extra thick layer of old leaves or straw on the henhouse floor. This first thick layer is important to the next clean out, as it keeps droppings from sticking directly to the floor, where they would adhere like glue. The henhouse furniture is replaced, new straw put in the nest, and I transfer the young chicks into the house. Then I get in with them and chase them around a while. This isn't to be mean, but to find where I must adjust the litter so they can't all run into a corner, smothering half their number. Chickens are truly "chicken," and till they get used to the adult birds, the young will all run in terror into a corner or under anything handy. They pack in on one another and the birds on the bottom will placidly allow themselves to be smothered. We've lost up to half of a batch of chicks that way. But by piling the litter up high in the corners and cramming it tight into any crannies, the smothering—or most of it—is prevented.

For the first two days, so long as the weather is decent, I leave the sliding door open enough so the small birds can peek out,

Here I am "harvesting" one dividend of poultry raising, a foot or more of droppings and old litter that will guarantee a bumper crop of garden vegetables.

but the adults can't get in. A few eggs are lost and feed and water must be provided separately in the run, and if it rains, I have to put in a few old wooden crates to keep the older birds dry. But in a few days the youngsters will get up the courage to come out and it's safe to let the two groups mingle.

There will be a lot of chasing of chicks by the rooster and top hens, but the crowding that can cause smothering comes only at roosting time, just at dusk. For the first few days, I try to be at the house when the large birds all come in to roost. If the little ones are scared enough to run to a corner, I just poke in a long stick and stir them around enough so they aren't all in one place.

YOUNG AND OLD MEET

At any other time of year the chickens are pretty much self-sustaining except for keeping up with food and water and replenishing the top layer of litter from time to time. The litter is nature's deodorizer and so long as a loose layer of old leaves, straw, wood chips, or whatever other organic matter is handy is kept on the house floor, there is simply no objectionable smell. The outside run is treated to frequent layers of litter also, particularly in the spring when thawing ice and snow and/or spring rains make the run surface a thick mud. In addition to

Come the rains in spring and fall we are glad we've saved a good supply of maple leaves to serve as odor-free chicken pen litter now and garden-richening compost later on.

CHICKEN FEED

droppings, the run has a considerable garbage content; all our kitchen, garden, and general grounds wastes are shared between one or another animal and the chickens get their share. In particular, they receive all calcium-rich egg shells and clam shells (smashed up fine), plus coffee grounds, regular and herbal tea leaves, and citrus rinds. The other animals don't much like these things, but the chickens think they're great. What they don't eat gets stirred into the litter. It either dries or composts quickly in dry weather, but in spring, only a thick layer of old maple leaves keeps the place from smelling like a garbage dump. (The recollection of that first spring when we had chickens, but no leaves collected, is enough to get us raking and bagging the several dozen feed sacks of maple leaves each fall to sweeten up the chicken pen each April and May.)

I'll discuss the dry feeds we grow and mix in a later chapter; for now suffice it to say that it took several years of giving

the chickens all bought feed before we had the land tilled and
enough know how salted away to grow our own dry ration (and
frankly, we still have to buy feed now and then after a bad season).
But garbage provides the single most important chicken feed
ingredient, the portion that is most costly if purchased—the
animal protein. I suppose that in nature, the wild ancestors of
chickens ate bugs. Our birds eat what they can get, of course, but
that isn't much. We supply these proteins by giving the flock a
frequent intake of suet, meat scraps from the kitchen, or the scraps
and inedible innards of animals we've slaughtered. The suet is put
in a plastic mesh sack and hung from a pole supporting the net
over the top of the run. The birds have become used to it, and
know there is a real treat at hand when it's dropped in, so they
come running. The offal of slaughtered animals (which can
amount to more than a quarter of a hog, say) is cleaned of any
excrement, then run through the big meat grinder/sausage maker.
It's mixed with a coarse-ground grain for a binder, wrapped with
twine to hang from later, and frozen in two-pound chunks. When
one of these is hung in the pen the chickens peck it clean as fast as
it thaws.

Sex and Other Complications

By the time they are about a month old, young males will
show a somewhat larger comb than the pullets. In a few more
weeks they will all begin working their way into the adult
"society" of the chicken pen. Young cockerels will strut around,
occasionally having brief squabbles with one another and
eventually taking on the older roosters. The young hens too will
chase one another around. All this is to establish the "pecking
order," which is more than a saying. It means, literally, that the
birds set up a status system. The top hen can peck any "lady" in
the pen, but won't be pecked. No. 2 can peck all but No. 1, and so
on down to the lowest bird, who gets pecked by everyone. Roosters
pretend to be above it all, but they get henpecked by just about any
of the females.

THE PECKING ORDER

This "social pecking" doesn't amount to much. But
pecking of a different order can come about if birds are over
crowded or if they lack fiber in the diet, and take to eating one
another's feathers to help digestion. Once a bird begins to bleed,
even a drop, the others are attracted to it and will occasionally peck

it to death. When a sore appears, I remove the beleaguered bird, dab some purple gentian violet "horse liniment" on the wound, and segregate the bird till the sore heals and feathers begin to cover the bare skin.

MATING   Bad pecking usually occurs when I've let a batch of young cockerels go long enough to figure out the facts of chicken life and begin to notice the females. Chicken mating is a pretty haphazard matter. The male grabs the female's comb in his beak, sinks his claws into the feathers in her back, and then as she runs around flapping and squawking, he tries to get his vent near hers to spray out some seminal fluid. With a fair amount of luck, some of the sperm will find its way into the hen's reproductive tract, migrate up the appropriate canals, and lodge in a special organ, there to fertilize each egg as it is being produced.

Nature has compensated for the hit-and-miss nature of this process by increasing its frequency to the point that it's going on most of the time at one corner or another of the run, if the rooster population is high. With either the scrabbling between cockerels or the battle of the sexes going on all the time, enough feathers can be lost and skin scratched that blood-pecking can become serious. Once an entire flock gets into the pecking habit it is just about impossible to stop them. Commercial operators de-beak their birds so they can peck but not hurt anything. Of course, they can't eat anything but premilled mash either.

CAPONIZING   Time was, when pecking started we had to reduce the flock by about half, even if the young cockerels hadn't reached full slaughter weight. Just last year we pretty well solved pecking and greatly improved the quality of our meat birds by getting a caponizing outfit. (Write to Howard Beuoy, Cedar Vale, Kansas for prices.) Caponizing is major surgery to remove the male sex glands that are located in a young cockerel's abdomen. I won't go into details because you have to get a kit, unless you have access to a surgeon's bag. The kits come with complete instructions. The only advice I'd give is not to waste money on the electric caponizer gadgets. You still have to make the incisions in each side of the bird's chest. And if you do go in for caponizing, don't worry that you are harming the animal. They don't appear to mind, and I've never had a single one show any bad effects from the operation.

When the young capons reach slaughter weight—often half again as large as they'd be if left unaltered—and we crank up the slaughtering equipment, I also cull the laying flock to remove unproductive hens. The books tell you all sorts of rules about eye

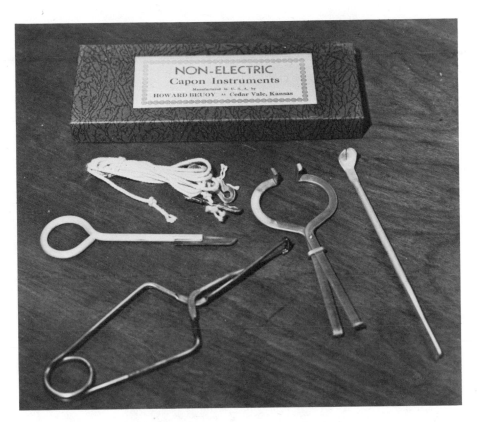

The caponizing set used to perform the operation [major surgery, really] that turns a young rooster into a tender-eating bird that can top ten pounds.

brightness and walking stance and feather conditions to tell good laying hens from bad, but such rules apply only if all the birds are the same age. Ours are anywhere from four months to four years of age and I apply a single culling criterion. If the bird is old enough to be a layer, she'll have a fully-developed comb, a good weight (only experience can tell you this), and her vent will be large, whitish in color, and moist. She'll also be pretty feisty and complain like the devil about being grabbed. A too-light bird, one with a small, dry, or puckered vent, or any bird with legs bleached out white from egg laying is a likely candidate for the chicken or s.ew pot. If in doubt, I segregate a hen along with a rooster for a week in a wire cage inside the run, so she won't get lonely. Given a small box of hay she should keep on laying at her usual rate, if she's laying at all.

CULLING THE FLOCK

Diseases

PARALYSIS

We started with day-old chicks from a fine New England hatchery, and have never had any serious disease problems. Once in a while a bird will just lie down and stop walking. Paralysis is a relatively common but incurable disease. They say the birds can be eaten, if you kill them shortly after they lie down. We don't, but kill them and give them to the pigs. (Friend, if things like that bother your stomach, I'm sorry. But this is the real world out here, not a supermarket.) Occasionally a bird will become crop bound. The feed in its crop, a sort of storage sack in its neck, gets clogged and nothing will pass through. The first hint is that the bird gets so weak it can't walk. Nothing can be done; the animal is so wasted it's mostly skin and bone. Pigs get it too.

WORMS

Intestinal worms are present in nearly all natural soils, but they are a natural parasite of chickens and harmless if they don't get into small chicks. By not letting chicks onto soil till they are a month or more old, we avoid worm problems (or have to date). If I ever detected excessive concentrations of stomach or intestinal worms in any chickens slaughtered, I'd dose their water with a standard veterinary worming medicine and clean both pen and run extra well.

OTHER DISEASES

By keeping the chicken netting over most of the top of the run, we keep out wild birds that can carry cholera and other bird diseases. Just about the only other problem we've had is frostbite. In severe winter cold the tops of our roosters' combs freeze, turn black and fall off. It doesn't seem to bother them or their females in the slightest, so it doesn't bother us. They just look a bit odd having smooth-topped combs with the characteristic points frozen off.

Breeding Brooding Back In

GOING BROODY

I mentioned earlier that we are trying to get the nesting instinct bred back into our flock so they will be self-producing without the use of incubators and the brooder. So far, I can't claim much luck. Each spring our bantam hens all go "broody," which means they disappear somewhere for a few weeks, then one day appear by the feed bins with a clutch of fuzz balls in tow. This maternal instinct has been bred out of the larger hens, but each spring several will make fitful attempts at motherhood. They act fussy, clucking more than normal, and prefer to stay on or near a nest. If they persist in this nesting behavior, I take the two or three most broody and put them in a separate pen, each with a generous-sized nest box about a yard square, filled with fresh hay and kept dry on top by roofing paper and off the damp ground by a brick in each corner.

More often than not these hens will fuss around, fight over nests, lay an egg or two, and then forget all about it. If so, they go back into the pen and we continue eating or selling their eggs. To date we have had four hens that raised up a brood. The largest and sturdiest of their offspring were banded, and Louise and I are keeping careful track of how they produce and whether any of the young hens show the brooding instinct themselves. (So far none of the "second generation" has raised up a brood.) We'll keep trying, and hope someday to have a real homestead chicken flock that requires nothing but feed and water and someone to gather eggs and fryers.

Rabbits

In many ways rabbits are a lot less bother than chickens and if they laid eggs, I imagine they'd take over from poultry. Rabbits don't crow or even cluck; they are perfectly happy in small cages, are easier to kill, a lot easier to dress, and ten times easier to clean up after than any kind of birds. All their meat is similar to the finest breast meat of chicken, more tender, if anything, and with proper cooking is in a class with prime milk-fed European veal. And on top of it all they grow fur hides, which aren't too durable but look cute on a little girl's coat or muff and are fine inner linings in clothing where warmth is a prime consideration. So why doesn't everyone have rabbits?

Probably the answer is that most Americans think rabbits are all like the Disney movie character with cute wiggly noses, pink eyes, soft paws, and meticulous habits. As every little city child who visits us says, "But how can you kill those adorable, fuzzy...etc.?" I'll admit, the first time was hard. Ending the life of a fellow mammal is probably one of the toughest hurdles a new homesteader has to get over. But with rabbits as with every other kind of livestock, familiarity breeds—not contempt, but hard, factual know-how. And there is precious little that's adorable about rabbits once you get to know them. A mature New Zealand White weighs a good fifteen pounds, has needle-like claws, and a kick that can rip your arm open. Under that wiggly nose is a set of teeth and jaws that can kill a rival, chew through two inches of solid oak, or rip a wire mesh cage to shreds in the space of one night. And when properly fed and watered, they produce

Neighbors Bill Corey and Robin VanAlstyne with a pair of young bunnies that had been rejected by the mother rabbit. Robin volunteered to bottlefeed them, and as the photo shows, did a fine job, keeping them alive till they could eat grass.

phenomenal amounts of manure—prepelletized in 3/8-inch globes and superb garden fertilizer to be sure, but a rank-smelling nuisance if not removed frequently. Cute they are not. But a single good doe, bred four times a year, will produce young that can dress out to over 100 pounds of protein-rich meat, and do it on well under 100 pounds of grain and a dozen bales of good hay. Even if all feed were purchased, that would come to less than twenty cents a pound for the meat, and the square feet of skins, garden fertilizer, and by-products are thrown in for nothing. Certainly every homesteader should have rabbits. We got ours the second year.

Choosing a Breed

Rabbits come in three sizes: small, medium, and large. The small ones are used mainly for medical lab work, the large ones for show. For meat the medium breeds such as the albino New Zealand Whites are generally considered best. Since this type is by far the most common, being kept by someone in just about every rural or suburban town, they should be easy to come by. The best way to get rabbits is to buy a pregnant doe and use her litter to start. Breeding males and females from the same litter apparently does no harm to these already highly inbred animals, contrary as that is to nature. Or you can start with a basic breeding mini-herd, a buck and two does, as we did. That will give you eating rabbits a couple of months sooner. Just don't buy the first rabbits around Easter, which is the only time of year the price is high. You should expect to pay about a half-dollar a pound live weight for young animals, perhaps a dollar a pound for a pregnant doe.

GETTING STARTED

I suppose the most natural way to raise rabbits would be to let them hop around in a large yard. The first problem would be a sudden bunny population explosion. A doe kindles (that's rabbit talk for giving birth) eight young four times a year. Half are females that can reproduce in five months or so. That's sixteen more does kindling six times a year—get the idea? The second difficulty is keeping the animals in, no matter how strong the fence. Domestic rabbits are descended from European stock. And unlike our American rabbits, they are great burrowers. Where our little native cottontail or the northern varying hare make a nest under a fallen log, the domestic rabbit mother-to-be will tunnel down as much as five feet. Then she will tunnel right back up again—on the other side of your fence. And finally, unless handled enough to become tame, a pet, and difficult to kill, they will flee humans in a most undomestic manner. If on rabbit-slaughtering day you chase them around the pen for any time they will get excited and won't bleed out fully, thus reducing the quality of the meat.

ONE "DON'T"

Equipment

So we keep the adult rabbits in individual sections of a three-section cage located along the shed wall just across from the chicken pen. The outside portion or run of this installation was designed to fit under a back-sloping roof made from a standard piece of four-by-eight-foot exterior-grade 3/8-inch plywood, which has roofing paper stapled on it for further waterproofing. The cage bottom is about a yard off the ground on stout legs, the bottom and sides framed in one-by-three-inch scrap lumber. The two dividers are of thin plywood, kept in place by bent nails hammered into the front and back framing so they can be removed to make a single triple-sized cage or a double at either end.

THE CAGE   The floors of traditional rabbit cages were made of plywood strips with a half-inch space between them to let the droppings fall through. However, such a floor soaks up urine and can be a breeding place for disease. Modern cages have wire cloth floors: steel wire woven in 3/8-inch squares and galvanized. The cloth is quite strong, but it must be fastened very securely to the wood framing, as we discovered when our Shepherd liberated our first three breeders by unravelling the cage bottom from one side. When I replaced both the rabbits and the cage bottom, I also nailed enough old boards to the legs across front and sides to keep dogs and other unappreciated rabbit-lovers out. It's a simple matter of pulling four nails to clean out under the pens.

Originally the sides and back of the pens were chicken wire. But when an amorous buck ripped the back out in his romantic pursuit, I replaced the mesh with stout hardware cloth. Each of the three sections has a door—hardware cloth framed in wood—that rests on two nails in the bottom frame and is kept in place by wood blocks that revolve on wood screws. The first couple of winters this was all the pen we had. I put old feed sacks at the north end of the cage to break the wind, and the rabbits did fine even when the temperature dropped to well below zero. Feeding and watering was a problem in deep snow, and we now have an inside section to each cage.

In the shed wall in back of each of the three sections I sawed a round hole six inches in diameter, with the bottom about three inches above the pen floor. The wire cloth at the pen back was snipped in an "X" pattern and the pieces pulled through the

Two breeders in the three-section cage. The twice-stapled chicken net, board across the front bottom and cage doors are all security measures taught us by our own dogs, who like rabbit too.

hole and stapled to the shed wall. Then along the wall, I built a box eight feet long and about a foot high and deep. It is supported at the front on stout log legs, the inside can be divided into three sections with dividers that slide up and down in wood channels, and the top is in three sections that hinge up from the back. The floor of the box is more wire cloth, and tacked under each of the three sections is a feed sack to catch spilled feed and droppings. Three or four times a year I remove the sacks and haul to the gardens or a compost pile a hundred pounds or more of air-dried, or in winter freeze-dried, prepelletized, prebagged organic fertilizer that assays out to about 3 percent nitrogen, 2 percent phosphorus, and 1 percent potassium. That's the equivalent each year of a 100-pound bag of commercial 10-6-3 chemical fertilizer that would cost more cash than we spend on the rabbits in a year. Now, if that isn't an argument for living on an organic homestead, what is? For the price of a bag of chemicals that would only run off the land to pollute some stream, plus a few hours of honest physical labor, you can have the same amount of plant nutrients, but in organic form that will become available to the plants gradually while adding to the tilth of the soil. And then as a bonus you also get a meal of fried rabbit once a week and enough skins to make two warm robes. I may have said it before, but why doesn't everyone keep rabbits?

The newest litter of rabbits is turned into the pen in back and I clean up under the junior cage—two wheelbarrow loads of nitrogen-rich droppings that will go between the corn rows.

Feed and Maintenance

For the first few years we were more or less dependent on agribusiness for rabbit feed. They have simplified the process by manufacturing a complete ration in pellet form. You can buy it in fifty-pound paper bags that zip open easily, for about five dollars. At least half of the cost is for convenience, the pretty picture of a bunny on the front, and the fiber content—since rabbits, like most grazing animals, require a lot of bulk in their diet. After buying a few bags of the stuff, it occurred to us that it made a lot more sense to provide the fiber naturally, with grass or hay instead of money.

A brief investigation proved that all a rabbit really needs is all the hay it can eat, all the water it can drink once a day, and a good cup of grain fed in three or four portions over the week. A bunnylick of salt mixed with needed minerals and frequent treats of stem, stalk, or leaves of anything green complete the ration. Pregnant and nursing does need somewhat more—we give them all the grain they want, which amounts to a third- to a half-cup a day, and keep fresh water available at all times. At first the grain supplement used was pig feed, about two-thirds the price of the rabbit pellets, and used in much smaller quantities since the main feed was hay. Now the rabbits get the same dry grain ration we feed the chickens. You can use bought feeders that hang on the cage—we just use ceramic bowls heavy enough so the animals can't tip them.

FEED NEEDS

Picking up a rabbit

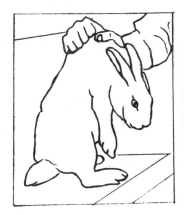

All during spring, summer and fall, and in the early presnow winter months, my morning exercise is cutting large handfuls of meadow grass for all the animals. The rabbits in particular relish it, and both their grain and water requirements are much less than in winter. I don't doubt that you could raise rabbits on nothing but the naturally-occurring feed, perhaps adding a few odd twigs and tree leaves and grasses with matured seed heads to the green grass. The animals would probably be smaller, less healthy, and surely would have fewer and smaller litters. We know they'd survive for a fact because of Wild Hare, the only rabbit that stayed around the place without getting dog-caught after we learned the hard way how rabbits burrow out of pens set on the ground. Ol' Wild Hare has been around for several years now, occasionally sharing some dog food or cleaning up after the pigs, but he (or she) lives mainly on wild stuff, summer through winter. The dogs have taken to treating it as a playmate, chasing it at a slow speed, not barking as they usually do with wild rabbits or other domestics that get loose on occasion. Wild Hare has enough holes and burrows around so it can escape an attack from any quarter. If it's a female, maybe it will appear one day with a litter that's half cottontail and we'll have a new strain of rabbits.

Breeding

In nature rabbits owe their survival to their amazing fecundity. A favorite and fairly easy prey of every meat eater from the tiny shrew to human hunters, rabbits continue to exist only by reproducing at a tremendous rate. A cottontail doe in a warm climate can produce a litter of six to eight, six times a year. On average, only one bunny per litter lives to grow up, but without Nature's population control measures, that number would still have the world covered with rabbits in a year's time.

Our domestic rabbits are technically capable of reproducing as frequently as their wild cousins, but it isn't good for them. Once when the animals were still teaching me the basics of cage construction, the buck broke down a separator and impregnated a doe that was nursing a litter only two weeks old. Thirty days later she bore her accustomed nine young, but since she had just weaned the earlier litter, she hadn't the milk to feed the new batch. As usually happens when a doe can't care for her young, she killed them all and ate several. As I said earlier, there's nothing particularly cute about rabbits.

HOW OFTEN   Some books recommend leaving a doe with her litter for the full two months it takes them to reach fryer size. Then a rest period of two months is supposed to follow before rebreeding. This would give only two litters a year and I think it's going too far in the opposite direction from Nature's six litters a year. We go on a four-month schedule, breeding a doe, say, on January 1, letting her nurse the brood for the month of February, resting her for March and April, and rebreeding on May first. A schedule of breeding two does in this way is shown.

We could fit in an extra litter per doe each year, but this would require a whole new set of cages. Note from the chart that there is a new litter coming on every two months. All birthing and litter raising is done in a single "junior cage," about a yard square with a hinged top, which makes tending easier than a latched door, and an old nail keg on the wire cloth floor inside. Exactly twenty-seven days after she is bred, I put a pregnant doe into the junior cage. The litter that has been occupying it goes into the freezer. The doe kindles on the thirtieth or thirty-first day and nurses the litter for a month. The day she is taken from the litter, I breed the other doe. When she is ready to go into the junior cage, the previous litter is eight weeks old and ready for the frying pan. And on it goes.

With such a schedule, you would think that rabbit breeding is practically automatic, and it is. Unlike most mammals, ovulation—production of the egg or ovum that will unite with the male sperm to produce the young—is caused by the act of mating. Once or twice we've had a mating that didn't take, but it is extremely rare, and if it happens more than once with the same animal, it's time for a replacement.

To breed a doe, I reach slowly into her cage, take her   MATING gently but firmly by the loose skin in the small of her back, and take her to the buck's cage. After being kicked and gouged severely several times, I now wear stout horsehide gloves with gauntlets and am sure to grab the animal from the rear, with my wrist over its hind quarters.

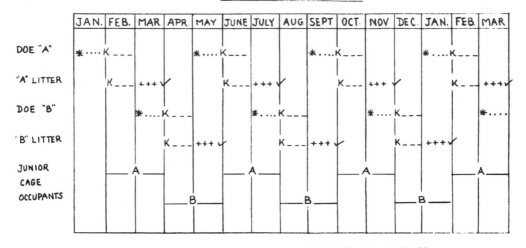

RABBIT BREEDING SCHEDULE

LEGEND: ✳ BREEDING
.... GESTATION
K KINDLING
--- NURSING
+++ FULL FEEDING OF YOUNG
↙ SLAUGHTER

The mating should take place immediately if the doe has not been excited enough to begin kicking during the transfer. After some preliminary sniffing and perhaps a brief run around the cage, the male will mount the female from behind, make a series of rapid movements of his rear end, typically keeping it up till the doe gets bored and either hops away or sends him flying with a kick of her hind legs. I've seen does kick the buck off after just a second or two and treat all further advances with a low growl. I used to think it necessary to remove a cage partition in such cases to let the pair work out their marital problems for a few days. But it apparently isn't. The doe knows her business, and brief affair or not, she usually kindles almost thirty days to the hour from mating.

There is always the danger of a fight in leaving two adult rabbits in the same cage for long. Bucks will fight just because they are bucks, and as she nears term a pregnant doe can become mean. It's best to keep adults separated even though at times they may appear to want to join the rabbit in the adjoining cage.

Birthing Time

During her final week of pregnancy the doe needs more water than normal (she has been on the daily grain ration for the full month) and will benefit from a daily offering of fresh greens. On about the twenty-fifth day, she will begin carrying hay around in her mouth, looking for a place to build a nest. Before the pregnant doe is transferred to the junior pen, I clean it out well, removing all old bedding and packed manure in the nail-keg OB ward, and washing off all feed and water crocks. Then she goes in along with a good armful of straw or old hay. She will pack the hay into the nest box and will pull out enough of her own hair to make a warm, cozy nest about the shape of a half grapefruit.

BIRTHING   The young are usually born at night over a several hour period. If the doe goes out of the nest for food or water during the birthing, the bunny will die of cold unless it is a very warm summer night. To minimize that problem, on the twenty-ninth night and as long after as needed, I shove the feed and water up to the nest at dusk. A check in the morning (with the heavy gloves, as a new mother rabbit will bite) usually shows eight good young rabbits, and one or two runts, deformed, or dead. These, plus the chunks of afterbirth the mother hasn't cleaned up, are removed. I make sure there is enough fur in the nest to cover the young adequately. If not, I pull some from the mother's ruff, and she gets a couple of carrots for her labors.

Newborn bunnies are a raw pink color, hairless, blind, and helpless. But in a couple of days they begin to grow white fuzz, in about ten days their eyes open, and in another few days they are out of the nest, hopping around with the mother. This is a dangerous time to invite a family with young kids to come over for a rabbit dinner, because, dog-gone-it, the tiny bunnies are cute. If a visiting child simply must have a bunny to take home, it's OK so long as the small baby bunny droppings have appeared in the cage. If the mother is still cleaning up after the young, it's too early to move them.

Incidentally, we find that gift-giving is the best way to dispose of old breeders. After three years, the conception rate and number in a litter will fall and both bucks and does should be replaced. By that time they have become pretty tame and easy to handle and we'd rather see them spend another five to ten years as someone's pet than be served up as a chewy dinner. Females are replaced with the largest young doe we have and bucks with a young male who shows his potential aggressiveness by exhibiting the appropriate precocity in the junior cage.

Telling males from females is easy, by the way. In a mixed group of ''adolescents'' more than three or four months old, the females will have their coats tinged on top—colored by the constant affectionate attentions of the young males. Otherwise, you have to inspect their privates. Both sexes have a small tuft of hair, a furry pad with a small opening in it, just under the tail. The sexes look quite similar at first, but a matured male will have two small half-rounded lumps flat against his stomach on each side and forward of the pad. These are his testicles. The penis can also be made to protrude slightly from the body by pulling the tail back and the front part of the pad forward slightly.

Figuring the sex is important for any youngster to be kept past the two-month age. They should not be bred till at least four months of age and six is better for their health. They won't go along with that idea if left to their own devices, however, so the sexes should be separated at two months.

THE NEWBORN

Carrying a small rabbit

SEXING

Carrying a large rabbit

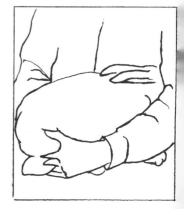

Problems

Rabbit diseases are mysterious things for which no one has bothered to come up with cures. The animals seldom get ill and if you lose one rabbit it's easy·to get another. We have only had one fatal rabbit disease and it came in with a young bunny we loaned out to a child for a while. It ended up a minor plague, killing a whole young litter and two adults. Afterwards all hutches got a special cleaning and disinfecting and we've had no recurrence.

Hosing down the rabbit Junior cage between litters. Keeping things cleaned up practically guarantees an absence of disease.

Sometimes a young bunny's eyes get gummy so the lids stick together, but a rinse of warm water will cure that. Life in the nest is also a pretty brutal affair; the blind young instinctively kick each time the mother enters, and the weakest young can be kicked out unless the lower half of the nest box is closed well. We learned this the hard way after a doe had chewed the bottom off. The litter kept kicking one another out, and once a bunny is out the doe will not feed it or attempt to put it back. We lost two young before I realized what the problem was.

SANITATION

Frequent removal of droppings and replacement of old bedding is an important sanitary measure, and I routinely disinfect the adult cages twice yearly and the junior cage each time a new litter is due—with Chlorox and hot water only. About the only other health measure I can think of is never to re-pen a rabbit that has been loose long enough to get out of sight. It may have come in contact with a wild relative that was a carrier of tularemia (rabbit fever), which is dangerous to humans as well as bunnies. In this vein, any animal that acts listless and stops eating for more than a day or two should be killed. If its liver has small whitish spots on it, have a vet or the local Public Health Service check the carcass. If it's tularemia, you'll probably have to destroy your stock and begin over. The authorities will know what to do.

Actually, the chances of domestic rabbits contracting this disease without direct contact with a wild bunny are about nil. So keep yours well caged and well fed and enjoy some fine eating for a minimum of time and effort.

6 Spring and Summer in the Fruitlands

If there is anything repeatable in mixed company to be said about New England weather, it's that it fits folks to the climate. After nearly six months of being cooped up by ice and snow, spring and early summer finds us with double the energy. Which is a good thing, since there is about ten times as much to do, not the least being early chores in the fruit plantings. The first due date is May 15, the average last frost with any teeth in it. After that, blossoms begin to pop and it is too late to perform several important chores. The first is to assure we have a good supply of our partners in fruit culture, nature's pollinators, the bees.

Bees Before Blossoms

I wait till we've had a good week of fairly warm weather in April, or sometimes not till early May, to remove the hay covering around the hives and take each one down. The objects are several: to see if the colony has survived well, whether they have honey and pollen to last till the spring flowers bloom, and to see if the queen is alive and laying. You can tell at a glance if the bees are alive and well, even if they are still gathered in the winter clusters in the top frames. No problem with stores either; if there are not four or five good frames worth of unclaimed stores, I plan to feed sugar water and a commercial pollen substitute in a special pollen-feeding frame till the onset of honey flow.

Finding the queen takes some looking. If the weather is warm enough, the bees should be moving all over the frames, and somewhere in their midst is the queen, hopefully laying eggs. She is the longest bee in the hive, and the others make way for her. Often you can find her by squinting your eyes. The otherwise chaotic pattern of bee wiggle will form sort of a moving pattern as the queen scuttles through. Once I'm sure she is alive and well, I check for young just as when the colony was first started. If eggs and larvae are present, the laying in a regular pattern with no cells being missed, I close up the hive. If there are any problems, it's necessary to send South for a new queen.

LOCATING THE QUEEN

So far, Louise and I have never had to deal with a bad queen, though one winter, apparently, the queen of one colony died in midwinter when there were no young larvae or eggs. Unable to raise up a new queen, the colony would have been doomed. But after noticing the absence of queen, eggs, or young in the nest on the first spring check, I had a new queen air-mailed up from Georgia. Introduced the same as when the colony was started, she was well on her way to revitalizing the hive within a week's time.

Our hive.

Besides a couple of early checks just to make sure the bees are producing brood, the only other beekeeping task between the spring check and early summer honey-collecting time is to try and prevent swarming. This is the bee's natural method of expanding. They raise up a batch of young queens to take over the colony, then the old queen and about two-thirds of the worker bees take off for a new territory. The old colony is left so underbee'd that it is unable to produce any surplus honey—the honey that we collect for the morning toast. So it's important to prevent swarming if possible.

Each week from late May through June, I put on a bug veil and make a swarming check of each hive. No one knows for certain just what triggers a swarm, but overpopulation (too much new brood emerging too fast), as well as many climatic factors, are important influences. But some years some colonies will begin to construct "swarming cells," usually along the bottom and side edges of the frames. Cells just a bit larger than normal, with unusually bulged caps, contain drones. Others, less numerous, have a long downward-hanging cap that looks like a peanut. These contain growing queens. Smashing the queen cells is not a good

SWARMING

idea. For one thing, they may not be swarm cells at all, but evidence that the old queen is being replaced for reasons that only the bees know. More likely, the process is far enough along that the swarm will take place anyway with the old queen and it will be just that much longer for the colony to raise up a new queen.

The books suggest several ways to help prevent a swarm, but most are useful only in a large apiary where someone is working the hives constantly. For example, you can clip the queen's wings. When the swarm emerges, she can't fly, and falls into the grass in front of the hive. She could die unless the beekeeper sees the swarm, captures the queen, and puts her in another hive. Presumably the swarm will return when it realizes it is queenless, and then can be captured and reunited with the

SWARM PREVENTION

queen.

In a small operation, the best way to prevent swarming is to give the bees plenty of room, at least two full-sized hive bodies for brood raising and plenty of supers, the special frames and hives for bees to store the surplus honey. Even if the honey flow is several weeks away, when I find on inspection that most frames are well covered with bees, I put two shallow supers over an excluder on top of the brood frames just to give the bees maneuvering room.

The one swarm we've lost to date can be blamed on the post office. Some extra hive bodies I'd ordered were delayed several days in the mail. We were enjoying the spring sun in the back yard when we heard this peculiar hum that kept getting louder and louder. We thought it was some strange aircraft when Louise spotted the swarm, a swirling column of bees, perhaps ten feet across and twenty feet high, rising slowly from a hive and into the trees. I tried following them, but lost the swarm very quickly. I'll admit that it was one of the most spectacular sights Nature has ever offered us and we are both glad to have seen it. Still, I'd rather have had the honey.

One way to practically guarantee a swarm is to try producing section honey. In the special section-honey supers, the comb is divided up by little wooden squares. If the bees are cooperative, they'll fill the squares with comb and honey and you can snap them apart and sell each square as bee-packed sections of honeycomb. However, this isn't the way bees like to work, and all the little wooden barriers get them feeling crowded. Without a lot of manipulating by the beekeeper, they are liable to swarm, so we don't risk it. To make squares of honeycomb, I just cut up regular frames. It's a little gooey is all.

To prevent swarms, you have to keep the bees happily
working during the May-June swarming season. I do this with the
so-called "Demaree" plan. As near to May 21 as possible (that's　DEMAREE PLAN
the day before apple blossom time), I tear down each have a final
time. All frames of sealed brood—cells covered with dark yellow
capping wax and containing young that will emerge over the
coming few weeks—I put into the upper brood chamber. These
bees will be housekeepers, cleaning and building cells and storing
honey during the main spring honey flow. An excluder goes
between the two brood chambers, and into the bottom one I put all
empty frames and all those with grubs and eggs. I'm sure to put
the queen in the lower story too. Any frame that is mainly honey
(with white cell caps) is removed and replaced with a frame of
empty cells or a frame with a wired foundation. With this plan the
queen is kept happy with one section to fill with brood. The young
housekeeper bees have two supers in which to draw comb and an
entire brood-free, full-sized hive body in which to store honey, so
they and the flying bees are kept busy and happy too. An
interesting note from the books: when the flying bees come in
from the field, they transfer their nectar to a young housekeeper.
She sort of gargles the nectar for a while, evaporating excess water
till the nectar turns to honey. No wonder honey is better than
sugar for babies and folks with stomach problems—it's already
partially digested by the bees.

On to the Orchard

Once the bees' cooperation is assured, we can turn our
attention to the fruiting plants. We have a big strawberry patch
and a hill covered with fancy vine, tree, and cane fruits. But the
real pride of the orchard is a collection of apples, grapes, and
berries we have claimed (or reclaimed) from the wild. This
combination of domestic and wild fruits assures us that there will
always be good crops of at least one fruit each part of the year, no
matter how the weather behaves. A wet spring that ruins the
cultivated strawberries makes wild blackberries superproductive.
The month of subzero cold that kills the peach buds turns out a
crop of half-wild crab apples so heavy the tree limbs break.

One year out of three or four we have crops of wild　WILD
strawberries that are worth picking for preserves, though any year　STRAWBERRIES

the tangy little fruits are a treat when eaten as found. Nature makes out the schedule and there is little we can do in the off years. But in the on years, when temperature and rainfall cooperate through the spring, the crop is considerable. I suspect we have increased the frequency of good wild strawberry harvests by running sheep in the orchard meadow where the berry plants grow rank. Like domestic strawberries, the wild ones put out a multitude of runners each year and they must become crowded. The sheep, with their close-grazing, root-pulling eating habits, do a good job of thinning the plants while their wastes increase the fertility of the soil. More on the sheep in a later chapter.

Even in good years the wild berries seem to appear in isolated patches, at different spots each season. The fruit hang quite close to the ground and it is easy to walk past a bumper crop without even noticing its presence. We prospect the meadows in advance, looking in early spring for the small white-petaled flowers with yellow centers that shoot up from the crowns with their spray of three-lobed, serrated-edged leaves. The flowers appear to carpet the land, then disappear in just a few days. Anywhere from a few more days to several weeks later the berries ripen—much smaller and seedier then their domestic cousins but stronger flavored and in some seasons, sweeter by far. Just a handful added to each batch of strawberry preserves gives a special wild tang, or so we think. And through the winter each slice of toast with strawberry jam or preserves carries with it recollections of hours spent on hands and knees, rifling through the meadow plants in search of the tiny crimson globes with the matchless flavor.

GARDEN STRAWBERRIES

We've had mixed results with regular garden strawberries over the years, as we tried out various berry varieties and cultural methods. First, we tried a cute idea (from OGF) and planted a strawberry/asparagus bed. I cultivated the plot deeply and dug out trenches about eighteen inches deep and six feet apart. Every three feet in the trenches I planted a foot-long leathery-rooted asparagus plant, the roots spread out in a fan and the tip of the crown a good six inches below ground level. As the asparagus tips grew I gradually filled the trenches till they were back at ground level. These were dormant, two-year old roots of the **Mary Washington** variety, by the way. I planted them as deep as I did so as to be able to cultivate the whole patch shallowly without harming the crowns. Planting shallow, with the crowns just under the ground, would work just as well, but cultivation and fall and spring soil preparation means more hand work.

Between the asparagus rows I put in strawberries, the plants a couple of feet apart, and the first year they grew beautifully. The next spring, the first year we could harvest either crop, we had the very few asparagus you should cut the first year after planting, and great crops of strawberries. But then, when we let the asparagus mature into yard-high fronds, they shaded the strawberry plants and made weed removal nearly impossible. That fall after frost I cut the asparagus fronds, weeded and remulched the berries after raking the plot to grub out all the weakened, old plants. Such treatment is supposed to rejuvenate strawberry plantings, but it didn't work for us. The shade must have been part of the problem, and weeds too. Plus a simple loss of vigor in the berry plants, due probably to nematodes. These microscopic **PROBLEMS** hairworms naturally build up in a strawberry planting and attack the roots unless you poison everything that lives in the soil with a chemical fumigant. Rather than do that, we just let the asparagus have the plot. The plants are gradually seeding the entire area. During the year I pull out any perennial grass tufts, and in fall the plot gets its shallow tilling. The asparagus crop from an initial twenty-five plants now supplies all we can eat fresh, just about every meal during the two-month-long cutting period. As with many vegetables, asparagus loses so much in freezing or canning that we prefer to enjoy it hugely when it is fresh, and mainly look forward to the next crop through the rest of the year.

After the strawberry/asparagus combination failed, we tried other clever ways of interplanting and keeping a patch producing for several years. None of them worked well enough to justify the time, so we've given up and put in a new planting each year.

Buying Strawberry Plants

An organic purist might object to buying commercially-raised berry plants. To assure their stock is free of the many viruses that attack strawberries, the nurseries raise parent plants in screen houses to keep disease-carrying bugs away, but they also do a lot of spraying in the fields. However, these plants are inbred, carefully developed from the tiny wild strawberry, and unless we want to limit our preserves and desserts to the infrequent good crops of purely wild berries, we have to give the tame plants the care they depend on. The plants are shipped when dormant. They are bare-rooted and most of the leaves are dead, so there is precious little residual poison that gets into our garden soil. And after spending a full year with us in good, organic cultivation, the fruit is at least as poison-free as the air most of us breathe every day, maybe more so.

PLANT
SUPPLIERS

I buy plants from W. F. Allen Company, one of the several excellent strawberry-growing specialists located in the Salisbury, Maryland area. One year, just for convenience' sake, we ordered berry plants from a big combination seed/nursery firm. It took two long-distance phone calls to get the plants at all; they arrived in June, a good two months late for prime planting; and it turned out they were mailed from Salisbury anyway. From then on, I've ordered direct from the grower, Allen in preference to his neighbors only because I started with him and have never been anything but pleased.

VARIETIES

This isn't to say that every berry variety from Salisbury or anywhere else will always satisfy. We've tried over a score of varieties; there are dozens, each with a different combination of flavor, size, climate preference, growing habit, and disease tolerance. All the better strawberry specialists do a fair job of describing the varieties in their mail-order catalogs, but every grower has to do a little experimenting to find the varieties that are best suited to his own growing conditions.

We started out with "everbearers," expensive varieties that supposedly produce two crops, a large one at regular berry time in June and another in the fall. We found the fruit hard, sour, and not worth the effort. We've also been disappointed so far with the super-fancy kinds of regular June-only bearing plants. Though we continue trying out a new variety or two each year, our main plantings are of two common types, **Catskill** and **Sparkle**. The former is a mid-season bearer, producing a first set of huge, wonderfully-flavored fruit in mid-June most years. It is resistant to a common tomato disease, Verticillium wilt. The **Sparkle** plants bear a week or two later from very rank growing plants; the berries are smaller than **Catskill**, but produced in great quantity. The plants are resistant to red stele, a serious root disease.

PLANTING

I order fifty plants of each variety for earliest April planting—the plants are oblivious to spring frosts. When they arrive, I unwrap the bundles (of twenty-five), wet the roots well, put them in plastic bags in the refrigerator, and wait for the rain to let up long enough to go planting. To set them in, I put the plants in a bucket of water and take it up to the gardens along with my trapper's trowel. This is one and a half pounds of tool steel, 1/8-inch thick, over a foot long, and three inches wide with a sharp, pointed end. I first discovered it when digging sets to trap furs, and it is the dandiest hand digger a gardener could ever want. It would take a sledge to bend it, unlike the flimsy trowels sold in stores; one long edge is sharpened so it can be used as a hand ax or sod knife; and its plain trowelling ability is unmatched.

The same poles used to set Long Reach Squares are laid out in five fifteen-foot rows, three feet apart and heading in the preferred north-south direction. Then I jab the trowel in the soil and wiggle it to make V shaped holes about six inches deep and a foot apart in the row. After trimming off dead leaves and roots over four inches in length, I set the plants in, the crown (a sort of nub dividing leaves from roots) just at ground level, roots well covered, and leaves exposed. Soil is firmed *very well* around the

Trapper's trowel

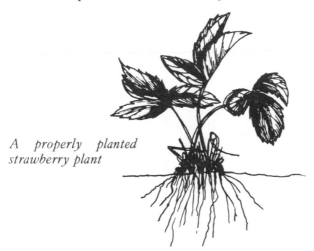

A properly planted strawberry plant

roots, and each plant gets a cup of water. I plant the two main varieties in alternating rows so they can share their different disease resistances. When all are in, the poles are removed and I forget the berry plants for a while.

Later in the spring, the exact date depending on the weather—after the first weeds are germinated but no more than an inch or so high—I run the trowel between the berry plants in the row, cultivate in the 'tween-row space, and then apply a thick layer of mulch that will keep in moisture and keep down weeds. After some experimenting we've settled on a two-layer

MULCHING

mulch that will last well for the year. First we shake out several feed sacks of maple leaves collected the fall before. The leaves are pretty dry and would blow off in the first high wind so, lacking a good rain that day (unusual), I'll wet them down and stomp them flat. The leaves form a fairly tight seal on the land, a perfect cover for earthworm activity. The other layer is old hay, a good six inches of it that's been fluffed up by a trip through the shredder.

Later Berry Care

PINCHING
BLOSSOMS

For maximum crops, strawberry plants must not be permitted to fruit the first year, but should spend their energy building strong roots. So, we pick all blossoms on a new planting as soon as they appear. (Of course, we're picking quarts of delicious berries from the planting made the spring before.) Later in the spring the plants will put out stolons, or runners, which will develop into a whole new generation of plants. I make two or three tours of the berry patch over the rest of spring, pulling any weeds that poke through the mulch and arranging the runners more or less in a butterfly wing pattern around each parent plant.

RUNNERS

I try to arrange runners and pull extra small plants so there is no more than one plant per square foot of space. This is commonly called the "matted row" system. It seems best for us, though I'm told that planting nursery plants closer together and picking all runners will produce more berries—but then you are buying five or six times the number of plants. As it is, we get up to fifty quarts of berries from ten dollars' worth of plants. With all-bought plants we might get sixty quarts, but the cost would be too high.

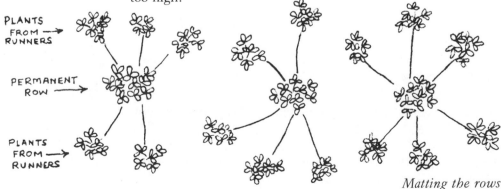

PLANTS
FROM
RUNNERS

PERMANENT
ROW

PLANTS
FROM
RUNNERS

Matting the rows

When fall arrives, the spring-planted berry rows receive a second thick layer of chopped old-hay mulch, heaped over the plants, to reduce heaving which could expose the roots during the alternate freeze and thaw of fall and spring. The mulch will be pulled back the next April, the plants will bear in June, and when the fruiting is done, the plants and mulch will be tilled into the soil and their place taken by a late planting of spinach or kale.

Wild Bush Berries

Thick stands of berries grow wild in several places around us, and often produce abundant crops when tamer fruits are hit hard by poor weather. The blackberry patches can quickly become masses of brambles, clogged with the accumulation of thorny canes that die each fall after fruiting, so that the fruit is practically unpickable. Each fall I take the Gravely garden tractor with a thirty-inch rotary mower/brush cutter we got (very) used, but still in perfect mechanical condition, and mow off one or another berry patch. (The Gravely is built precisely like one quarter of a four-cylinder auto engine, and though it's a hairy beast to hold in check, nothing short of a boulder or a two-inch-thick sapling can stop it.) After mowing, the berries in that particular location are lost for a year, but the following spring the roots bounce back with renewed vigor and produce large and easy-to-gather crops over the next three or four years, especially after a particularly wet spring.

WILD
BLACKBERRIES

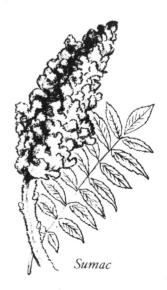

Sumac

Wild blackberries are pretty seedy, though the flavor is rich and strong. Louise often cooks up a big batch, adding just enough water to make steam, and lets the pulp drip overnight from the jelly bag. No seeds in blackberry jelly, which she makes by adding enough lemon juice to give some citrus tang, a cup of sugar (or half sugar and half honey) for each cup of juice, and cooking to jelly temperature, 220° on the candy thermometer.

My favorite way to use the blackberry juice is to combine it with sumac squeezings. About the same time the berry patch is purple with fruit, the red sumac trees are putting out their six- to eight-inch-long flowering spikes. The outer half-inch or so of the deep-crimson flower heads consists of small fuzzy globes that contain a citric-sour juice with a strong tang of bitter that I find pleasant to the taste, though everyone else in the family thinks it's awful.

Usually the inner cores of the spikes are wormy, so I pare off the outer layer and feed the cores to the goats, who adore them, worms and all. Then equal amounts of sumac parings and blackberries are boiled up in a little water and let drip in the jelly bag. I add just enough sugar to make the juice tartly sweet and boil it down to jelly temperature. I've read that you can substitute elderberries for the blackberries. Supposedly the eastern Indians got their vitamins that way. But years when the little umbrellas of elderblow turn out a bumper crop of shiny, black berries, we make elderberry wine.

WILD
BLUEBERRIES

A wild plant that we are in process of domesticating completely is the blueberry. The low bush blueberry, with a plant that reaches no more than three feet high, is common in several clearings near our place, especially where there has been a recent burn, accidental or from burning of logging slash. The plants are easy to spot in spring or summer with their crooked branches and shiny, oval, deep green leaves. We wait till late fall to dig them, keeping a good half-bushel of soil firmly around the roots, and are slowly transplanting one or two plants from each good stand to a portion of our hillside.

Blueberries demand a soil even more acid than ours is normally, so to plant I dig out a two-bushel-sized hole, pick out the rocks (about half the bulk), and replace them with a combination of white pine needles, oak leaves, and acorns raked up in the woods. All these contain plenty of natural acids. The berry plant is put in, soil packed well around, and on the soil under the branches goes a good layer of strawy manure from the goat pen, covered with a thick mulch of more pine needles and oak leaves. I can't say we have enough plants yet to provide all the blueberry muffins we could eat in a year, but we're getting there. And the small wild berries have a stronger, tangier flavor and firmer texture than the big pulpy hybrids you see in nursery catalogs.

The Vineyard

Early Viking explorers who reached New England named the new continent Vineland after the masses of wild grape vines growing thick in the forests. Most coastal woodlands, of course, have been replaced with housing developments or paved over with asphalt. But out here in our little backwater the maples, hickories, and evergreens still grow tall, and play host to vines that most every fall grow heavy with clusters of red and purple native grapes.

WILD GRAPES

It's safe to say that most wild grapes go to feed the birds, as well they ought. The plants require full sun, so in the woods, the vines send their woody trunks up to the tree tops. There they leaf out and bear fruit, hidden from the gaze of any human down below by all the tree limbs and leaves. In early fall when I'm getting the land in shape for winter, Louise prospects the woods for fall mushrooms, hickory, and walnuts—and grapevines. Each ropy vine gets a sharp pull, perhaps one time out of ten producing a shower of ripe grapes. Each wild vine is an individual, some with tiny, bitter fruit that are mainly seed, others sour, most with just a few berries. But occasionally one will prove to be a wonderful surprise, with great clusters of juicy, sweet grapes.

TAMING WILD GRAPES

If the vine is in dense woods, Louise just remembers the spot for future mid hike treats. But if there is a clearing within reach of the long trunk, and some are over fifty feet in height, we make a return visit to attempt "civilizing" the plant by pulling it down. Some vines refuse to cooperate; they are so well tangled in the trees they won't budge. They will be good for at least a generation of kids' games of Tarzan of the Apes. Those that do come down leave a lot of growth in the trees. This is a benefit, surprisingly. Wild grapes, just like their domesticated relatives, produce more and better crops if they are well pruned each year.

The final step is to build an arbor in the nearest sunny spot within reach of the vine. For each vine, I cut four stout saplings with a fork about ten feet off the ground. Cut and trimmed, they are set into two-foot-deep holes in a six by twelve rectangle. I fill the holes with rocks, and later with a concrete slurry if they are near enough to the house or a water supply that the hauling isn't a big chore. A framework of saplings is laid in the forks of the uprights, more poles are laid over them, and we weave the vine into its new home. If removal from its original tree hasn't done enough pruning, we cut the vine back, ideally so that only its four strongest tendrils remain. These are cut to no more than twelve feet long and all but a half-dozen of the pencil-thin old fruiting canes are removed from each one. Probably well over three quarters of each vine is pruned off, and the pruning is repeated each fall. But the vines respond dramatically, producing great clusters of fruits that are often twice the size of the earlier fully wild crop.

The wild purple grapes have too acid and "foxy" a flavor

Perhaps three years out of each five the wild grape vines we have "tamed" will produce the bumper crop Louise is admiring in this photo.

to make even a barely palatable dry dinner-type wine. The reds, perhaps with a bit of purple mixed in, can be coaxed into a fair dry wine, and both types make good dessert or aperitif sweet wines. Both kinds of grapes produce superb jellies, too. In a good grape year a single tamed vine has produced enough purple Concord-type grapes to supply our year's jelly needs and a goodly supply for several neighbors. As each year sees one or more new vines added to the family, our wine production increases.

Sissy Grapes

Other than their annual pruning, the wild grapes get no attention. However, the more delicate grafted fancy varieties need more care, since most are descended from vines developed in the balmier climates of France, northern California, or the Finger Lakes region of upper New York State. Indeed, it took several years before we found a bought grape vine that would survive our frosty climate, to say nothing of producing a crop. The superb blue variety **Alden** and the white seedless **Interlaken** managed to survive our -30° winters, but only from ground level down. Top growth was killed back each winter no matter how heavily I mulched.

So far we are having good luck with the **Delaware** red grape and **Worden** and **Van Duren** of the Concord family, but they are not all that different from the wild grapes. We have hopes for another good white seedless variety, **Himrod**, and a hardy wine grape called **Seibel 9549** by the Millers' Nursery, but the final results aren't in yet. Our objective in all this is to come up with a completely hardy grape that will produce a really good dry, red dinner wine. Since none of the professional hybridizers have been able to do it as yet, I really don't have much hope, but there's no harm in trying.

DOMESTICATED VARIETIES

Grape Planting and Care

Two-year-old grape vines when received from the nursery are puny looking things, requiring a planting hole no more than a foot across and deep. I dig out this cubic foot of soil, remove rocks, scatter a cup of bone meal on the bottom, then mix it with a couple of shovelfuls of peat, a little lime, a shovel of mature compost, plus enough soil to fill the hole. Roots are trimmed to a six-inch length, spread in a circle, and the soil mixture is firmed very well around them, the top left in a shallow dish-like depression to hold water. Like all newly-planted vines, bushes, and trees, the grapes receive a good watering each week of summer and fall that we have less than an inch of rain.

I locate the vines against the vertical supports on one of the twelve- to fifteen-foot arbors on the orchard hillside. They

PLANTING

have either a reclaimed native or one of the ranker-growing domestics growing at each end to provide arbor cover. The first year young vines need no training, but at planting are pruned of all but the single strongest cane, and this is cut off so that only the two base buds remain. The second spring, after hard frosts are done, but well before buds begin to swell, the vine is pruned again of all but the single best cane and this is tied with baling twine to two horizontal wires stretched between the arbor posts, one thirty inches from the ground, the other two feet above that. The third year the vine is a healthy adolescent five-year-old and we can expect a first small set of fruit. I choose the two best canes at each side of the main trunk and tie them out to the wires at each side. All others are cut off at the trunk but the two second-best (thickest) canes are cut nearest the wire levels. They are pruned to two buds apiece and will provide the fruiting canes for next year. In succeeding years the old fruiting canes are removed and the same new cane selection, training, and pruning as in year three is repeated.

Our vineyard.

PRUNING

In time the trunk around the wires will get pretty knotty, but that is normal and will not interfere with the vine's productivity. To get the most good grape clusters, many vines

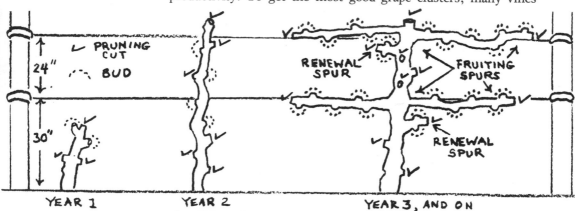

YEAR 1   YEAR 2   YEAR 3, AND ON

should have fruiting canes pruned to a certain number of fruiting buds—and it takes experience to tell just how many. A young or weak plant should only be allowed to produce on five or so buds per cane, and you go on up from there. Productivity is further increased by a constant heavy mulch/fertilizer applied each spring. I pile rocks around at the vine's base and each year dump on a good six-inch layer of soiled bedding from the goat pen.

We rely on the native plant's built-in vigor to carry them through disease and pest problems, and though there have been years when the crop failed, we have never lost a wild vine. The domestics are more vulnerable, though, and require attention, especially during the first two years when their foliage cover is thin and their root structures immature. Most years from mid-May into June we have a big hatch of rose chafers, skinny beetles with tan bodies, brown heads, and long, thorny legs. Hand picking for three or four days is usually effective. If not, I wrap the worst-chewed vines in plastic sheeting, leaving the top open so the vine won't toast. If the plastic "tube" thus created rises five or six feet above ground level the bugs won't try to fly over it.

Rose chafer

Our summers are seldom humid enough for mildews to be much of a problem and being on a hillside, the orchard air drains and flows constantly, which also keeps bacterial diseases from doing much damage. Still, I make a weekly inspection, removing all shriveled or wizened fruit and pulling, then composting or burning, any leaves with a powdery white deposit or black spots.

In late summer we have always had legions of Japanese beetles, the beautifully iridescent but voracious Oriental immigrant that tends to congregate in groups on the tops of large, flat leaves, and which sticks its prickly hind legs up at you when disturbed. In a bad year the beetles can ruin both foliage and fruit of grapes, cane berries as well as several vegetables. For a while we tried hand picking the bugs, jiggling leaves so the beetles would let go, then catching them in cans half full of water with a thin layer of kerosene on top. However, since they have a considerable flying range, we would just have a new crop each morning.

A few springs ago when we expected a really good grape harvest we swallowed hard and spent over sixty dollars for a ten-pound drum of DOOM, one brand name of the milky spore disease which government scientists found in the Japanese beetle populations in a part of New Jersey some years back. Ours came from one of the firms licensed by the U.S.D.A. to make the stuff, Fairfax Biological Control Laboratory, Clinton Corners, New York 12514. When spread by the spoonful in a four-by-four-foot grid in cropped or mowed sod and garden soil (where the beetles lay eggs) around the home place and orchard, the disease will attack the beetles in their grub stage. It will live on in the soil, being passed from generation to generation of beetle grubs, and will slowly expand out from our place wherever the beetles go. Milky spore will never completely eradicate the beetle, simply keep it in natural check so that it won't appear in huge numbers and do serious crop damage.

Japanese beetle

Even milky spore disease's most ardent fans admit that the effect is slow to take hold, and we still have more beetles than we'd like. But they do seem to become fewer each summer. Last year was the first time I didn't have to cover the best grape clusters with plastic bags till the Japanese beetles left. Perhaps next year they will be nothing more than an occasional curiosity.

If anyone is thinking that sixty dollars is a lot of folding money to spend getting a two-acre plot of sod rid of just one kind of bug, I might agree. But the cash and environmental savings are a lot greater than if we used whatever poisons they dream up to replace DDT over the next decade or so. And besides, I consider that money something of an investment in all our futures. Milky spore will spread out slowly from wherever it's applied and sooner or later the spread from ours will meet the spread from someone else's. The more folks use the stuff, the faster Japanese beetles will come under natural biological control, the better and easier will be

Armed with a glass of water and a quick wrist, Louise is about to rid the sunflowers of a day's accumulation of Japanese Beetles.

gardening for us all, and there will be just one less excuse for the chemical companies to claim their poisons are the only thing keeping the human race from starvation.

Orchard Fruit and Nut Trees

Our woods are full of hickory and butternut trees and the West Virginia branch of Louise's family sends up a box of black walnuts each fall. Though these wild nuts have a wonderful rich flavor, they are a flat bore to shell out. You have to dry the sticky outer husk, crack it off, break the granite-hard shells between a pair of bricks, and then if the inside isn't water-rotted or wormy or wizened, you spend tedious hours picking out tiny nibblings of nut meats. We relish the wildings, but save them for a special treat, often as not when there is a good supply of teen-agers around to do the shelling.

WILD NUTS

We did want somewhat more civilized nut crops for desserts and the holiday nut bowl. A little investigation revealed that pecans, almonds, and filberts (hazlenuts) would fare poorly in our climate. This left only Carpathian English-type walnuts to augment the wild nut supply. Since the wind-borne pollinating process needs two or more trees for success, we put in three Carpathians along the crest of the orchard hill. Originating in Siberia and distributed out of Canada by the University of Wisconsin, they promised to stand up well under our winters and they have. Only one tree so far has experienced any winterkill, and that was not serious enough to cause permanent damage.

NUT TREES

Being tap-rooted trees, putting down a huge, deep-probing central root, the Carpathians get started slowly, but reportedly put on four or five feet of growth a year once they are well established. So far ours are growing more in inches than in feet and I doubt that we can expect much of a crop till they are eight or ten years old. But the wait will be worth it. Nuts supposedly fall free from the husk, are easy to crack and shell, and a mature tree can be expected to give three or four bushels of shelled meats. That means meats, pure eating! Now three trees that cost fifteen dollars several years ago will give, say, ten bushels of nut meats. English walnuts are going for about a dollar a half-pound or cup, and if there are four cups to the quart, eight quarts to a peck, four pecks to a bushel—well, you figure it out if you want. For this sort of production, any nut tree needs a large amount of readily available food, potash in particular. We are letting their soil store it up for the future, giving each nut tree the mulch our young fruit trees receive plus a scattering of several shovelfuls of wood ashes each spring.

Restoring Old Apple Trees

We found a fruit equivalent of wild nut trees when we first wandered the woods of our homestead—ancient apple trees. I'd bet they were eighty or 100 years old, gnarled and loaded with dead wood and sucker growth, choked in by scrub trees so that some put scrawny limbs fifty feet into the air. And not a one was producing enough good fruit to make a glass of jelly. Our first spring in the fruitlands was spent doctoring these old-timers back into limited production.

SELECTING   THE
BEST

The first problem was how to tell the once-cultivated, now gone-wild trees from the hundreds of truly wild seedlings. Apples don't grow true from seed and a tree that jest growed is seldom good for much but rootstock to graft good-bearing scions onto. The problem was, many old settlers did just that—grafted fancier wood onto native seedlings. So our trees weren't laid out in a nice orchard, but grew more or less wherever a seed had sprouted. The fall before we had looked for trees that were odd shaped, lacking the single straight trunk that is an apple's natural growth habit but which is cut out by most pruners, old-time or modern. Then we chose by taste, nibbling the few puny apples growing high in the trees' crowns or lying on the ground. We figured we might be able to improve size and yield, but not taste. A bitter seedling apple isn't about to change no matter what treatment it gets.

PRUNING

First, I used a power chain saw to clear out all the brush around each tree. Then later in the year when they were dormant we went pruning. All the dead wood, hollow or badly diseased limbs were cut out. Overgrown trees lost crowns, and all interior limbs, crossing or rubbing branches were removed. We tried to leave each tree its four or five healthiest limbs, though several proved to have only one or two branches worth saving.

SCRAPING

Next we scraped each trunk and limb, chipping off the dark, mossy scales that bugs love to hide under. Where a broken

The first step in restoring an old apple tree is to cut out the unthrifty, scraggly growth with the chain saw. I must have cut out more than half of this old tree. Then Louise went over trunk and limbs with a scraper to remove scale and bark that can hide bugs and their eggs. Finally, the remaining limbs were pruned back by about half.

limb or woodpecker hold had let in rot, I dug out as much of the black, wet wood as I could reach and painted the inside with homemade tree wound dressing (thick roofing tar with enough kerosene on top to permit me to brush out a workable glop.) This concoction also went on scars where limbs were removed.

Next spring I plugged all the holes where rot had been removed with cement patches, scraped the branches till each took on a pink, youthful glow, doused each tree well with a miscible oil dormant spray, and sat back to see what would happen.

PATCHING

Such severe pruning is said to throw even young healthy trees out of bearing, so we didn't expect much that first year. We didn't get much either. The trees flowered sparsely, leafed out only at their tip ends, and seemed determined to spend the year making suckers. Each of these long, shiny whips was snipped off as it appeared, and the trees seemed to sulk through the summer. In the exceptionally dry fall that year several just gave up—lost their leaf cover and died. I suppose the "cure" was fatal; too much wood had been removed and the trees couldn't produce enough leaves to feed their trunk and root systems.

THE FOLLOWUP

Trees that made it through the fall and winter, though, began to perk up. The second year a creaky old yellow apple producer turned out a good two bushels, though the fruit was more worm than apple. As we continued with our scraping and dormant spraying, and began experimenting with the several organic pest control techniques we now apply to the entire orchard, production continued to increase. In three years time the old yellow apple was contributing our full year's requirement of apple sauce, and the various reds produced plenty of fall eating and, with careful inspection, enough worm-free fruit to keep us through much of the winter. I won't pretend that these old timers are on a par with new young trees in either quality or quantity of output. But for good fruit quick, doctoring old apple trees can't be beat.

The New Orchard

At first I planned to develop an all-new orchard using the old-time methods, by grafting bearing wood of the newest and best fruit varieties onto wild rootstocks. But a little reading changed my mind. In the past few decades the nurseries have developed

The reward: bushels of fruit from a tree that had been given up as worthless years ago.

both planting stocks and growing techniques that have revolutionized home fruit growing, and once again we gave in to modernity—to the tune of over seventy dollars.

SELECTION CRITERIA

This was a time when the green folding stuff was in particularly short supply, so a great deal of planning went into our selection. First consideration was hardiness. Our winters can get down to -30° and with a strong wind, the chill factor blows the bottom out of any thermometer. The second consideration was timing of production. We wanted a good mix of bearing times, ranging from the six- to eight-foot-high super-dwarfs that often produce a few fruit the first year after planting, to standard twenty-to thirty-foot sizes of a variety like **Northern Spy** that won't put out a bud till it's put in a good dozen summers.

ORCHARD LAYOUT

The final consideration was tree size and arrangement. Commercial orchards are all planted in neat, straight rows, which likely makes for easier picking and spraying. But nature never planted anything in neat, straight rows and Louise and I just naturally distrust people who tend to arrange things that way, so we spent hours drawing out an orchard plan that would look as though designed by nature. We had to fiddle around to assure all trees would get full sun; then the dwarfs had to be planted nearest the house so the standards wouldn't block our view of any blossoms. Many kinds of apples, pears, and apricot trees won't pollinate themselves, requiring two to tango, so to speak, so we couldn't separate them by too much distance. And finally, we had to locate the trees far enough from each other and from garden plots that there would be little or no competition for food or water.

OUR CHOICES

The varieties we chose are especially adapted to cold climates. Peaches are standards of the **Reliance**, developed by the University of New Hampshire. For a species that prefers the balmy winters of Georgia, this variety does remarkably well for us. Some very hard winters will reduce the fruit set, as will very late spring frosts. But most years there is plenty of fruit. A **Reliance** peach is no Georgia Belle, though. They are harder and less juicy than **Elberta** or **Hale-Havens**, but flavor and sweetness are fine. Best of all, their tough Yankee texture keeps them intact through freezing.

Our apricots, **Moongold** and **Sungold** dwarfs, were bred to survive severe cold by the University of Minnesota. We've just begun to get fruit since the first **Sungold** died after its third winter. It was killed by mice or moles whose underground burrowing pulled the soil away from its roots. I suspect they were attracted by over-wintering grubs, who must have been attracted themselves by the heavy application of bone meal I mixed with the planting soil—the last time I used that fertilizer in that way, though it is highly recommended by nurseries.

Apple Crazy

Where some families nibble on Cheez-E-Blobs, or whatever they call those yellow puff things that come in plastic sacks, we are all apple nuts. Each year I set in one or two new trees just to increase our variety. Most are apple types you never find in stores. Outside of the old reliable **McIntosh**, the supermarket offerings are running to blah-tasting varieties that are selected for their keeping and shipping qualities. In our opinion the widely-sold **Delicious** apples, red or yellow, fall into this category.

Our **Anokas** are out of the Midwest, an extremely stout and hardy little tree that bears moderate crops of a fair-to-middling red-striped apple, but bears them quickly. Our trees are planted where they won't be missed if cut down after the more choice varieties come into full mature bearing. Both the **Cortland** and **Macoun** are semidwarfed **McIntosh**-type apples. The former bears a little earlier than standard **Macs**, but is blander. We find it something of a disappointment. **Macoun** more than makes up for it, however. If there ever was a super-apple, this is it in our opinion, with twice the spice and tang of even a good **Mac**.

The **Red Rome** is an early-bearing long-keeper with fine quality, a deep red and thick skin. The **New Monroe** is in the same category, also ripening very late in the season. Our newest addition (as of this writing) is a **Spigold**, a newly developed cross which turns out red with yellow stripes and reportedly has given some of the tang of its parent, **Northern Spy**, to the other half of the team, **Golden Delicious**. We'll know ourselves in three or more years.

OLD-STYLE VARIETIES

The final apple category is one in which I can claim no success as yet, but plenty of enthusiasm—old-style apple varieties. There is a growing movement in many parts of the country to rescue the old varieties from oblivion even though they lack the vigor or quality of newer kinds. Miller Nurseries, among others is offering over a dozen of these trees, with names like **Tompkins County King** and **Winter Banana**. The latter originated in Indiana in the 1800s and supposedly has a banana-like aroma. And then there are **Pound Sweet, Sheepnose** (or **Black Gilliflower), Snow Apple, Strawberry, Maiden Blush, Irish Peach, Seek-No-Further,** and dozens more. I'd like to have them in our orchard just for the fine old names. A distant neighbor makes a hobby of going around and collecting old apple varieties.

When I get some time, I plan to try talking him out of some branches so I can try grafting my own. One good thing about homesteading, there's always something new to learn. If there were only more time to learn it!

PEARS
Our pear trees are two French varieties, **Beurre Bosc** and **Beurre D' Anjou**. They are fully hardy, producing a firm, indeed a hard-fleshed fruit which keeps exceedingly well if picked a bit green. They will ripen in a few days at room temperature after being taken from the cold cellar, and are also firm enough to freeze well.

CHERRIES
The cherry trees are the newly developed **North Star**, a descendant of Siberian stock. They are budded onto super-dwarfing rootstocks, so when fully mature will be little taller than any grade-schooler need reach. This makes picking of the full-sized, well-flavored fruit a snap, and also lets me throw a cover of Durex anti-bird (and anti-kid) mesh over them when the fruit is beginning to change color, which is the only way we could harvest that first cherry on our place. This bred-for-hardiness variety is a sour cherry, similar to the well-known **Montmorency**. The fruit is good eaten out of hand, not as sour as a lemon, but better in a well-sweetened pie or tart. We'd love to be able to raise sweet cherries such as the **Bing**, but the weather says no. Firmly.

PLUMS
Climate also prohibits us from having the super-juicy Japanese plums, but we did plant two more hardy European varieties, the purple **Fellenberg** and **Stanley**, both on standard-sized rootstocks. Neither has fruited yet, and in fact they haven't shown much inclination to grow. They appear to be the favorites of every bug in New England and have been severely set back each year by rose chafers or Japanese beetles or both. The milky spore is reducing the numbers of the latter pest and I keep looking for the deep sand which is the only place chafers can breed. There are a couple of good suspects, old horseshoe-tossing pits, near the plums and I should dig them out, till them in, or salt them down. To date we haven't decided whether we like playing horse shoes or eating plums the best.

Orchard Care

The most important part of orchard care is the attention given young trees on planting and during the first year or two. Give a young tree super-care when it's a baby and it will be super-productive and super-fast for the next hundred years or so.

First, when I receive young trees from the nursery, I immediately unwrap them and "heel them in," which means laying them down in a shady place, covering the roots and lower trunks with moist soil, and covering the rest with several layers of burlap. Trees are sold only in spring, the best time for planting, or fall when they are dormant for the winter, but exposure to wind or sun will still sap them of valuable moisture.

HEELING IN

Next I prepare planting holes. On one or two occasions Louise has had the foresight to wheedle me into doing the really wise thing, preparing planting holes the fall before. Whenever done, I remove a circle of sod and top soil at least two feet in diameter. Then I dig down two feet. On the bottom of the hole goes the sod, upside down, then the top soil. I remove the rocks from the subsoil and mix in at least four heaping shovelfuls of peat moss and another four of "super-dirt," a mixture of nitrogen-rich compost from earlier clean-outs of the chicken house, wood ash or potash, and rock phosphate (in place of the pest-attracting bone meal) for phosphorus.

THE PLANTING HOLE

Before planting the young tree, I prune off all roots longer than twelve inches, pull out any "bird's nests" of fine roots that might keep soil from pressing up against the main root branches, and cut off roots that are broken or badly scarred from the nursery handling. This root pruning, so long as it is not overdone, is highly beneficial for the young tree. The tiny feeder roots never survive transplanting, so all the roots can do the first year is absorb water. The tree must live off stored-up minerals, and the less excess root and top growth there is to feed, the better.

PLANTING

Next I mound soil in the bottom of the hole into a low cone so that when the tree goes in, the nob or slight swelling at the trunk/root join will be properly placed. This is the graft, where a fruiting wood scion was mated with the rooting stock, and it must be well underground—below soil level—in all *standard* sized trees, so the rootstock will not put out its own growth. *Dwarf*, and *semidwarfs*, on the other hand, would lose their small size if the standard scion wood were allowed to grow. So their grafts must be kept out of the soil, well above ground level.

To plant, I place the roots on the soil cone, spread them out in as near a circle as possible, and put in enough more soil to cover two inches deep. Then I stomp the soil down hard, add more soil, and stomp, and repeat till the tree is very firmly set in the ground. I try to leave a goodly dish in the surface to hold water.

The young tree's trunk is next wrapped loosely with old aluminum foil, which extends an inch or more below ground level and up to the first strong branch. The foil will prevent sunburn, a real threat the first two years or so. And it will not restrict growth at all.

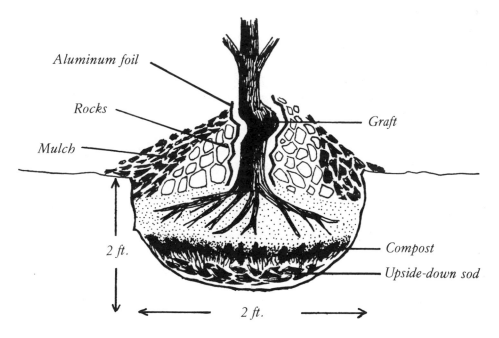

Next, I pile small rocks up in a cone around the tree's base, perhaps extending three or four inches above ground level, and put on six or more inches of leaves or old hay mulch in a donut shape, with the stone-surrounded trunk forming the "hole". This lets the trunk breathe even though it is foil-wrapped, and the stones are a further barrier to mice, rabbits, or sheep that might enjoy a nibble of fresh, young fruit tree bark.

Finally, each tree is given a thorough soaking with at least two three-gallon pails of water. This treatment is repeated each week during the trees' first summer and fall, unless there has been at least a full inch of rainfall.

Pruning

There are whole books written on the secrets of pruning fruit trees, but those so-called secrets are a mystery to me. Common sense and the barest understanding of nature's processes is all it takes. With a young tree just ripped out of the earth, you have to somehow make up for all the root structure that was lost by cutting away some of the top growth. If the grower doesn't do the pruning, the tree will do it alone. The problem is, the tree may not choose the best branches to kill off. So, after planting a young tree, I prune for two things: to get rid of excess top growth, and to shape the tree.

COMMON SENSE GUIDELINES

Since each kind of fruit and varieties within the same species, as well as individuals within each variety, will vary tremendously, there is no set rule to the first pruning. I try to leave the same number of top branches as there were main roots, since trees' top growth and root structures tend to be mirror images of one another. With a tree having just a puny two or three roots, I prune severely, no matter how vigorous the top looks. Even with well-rooted trees, I remove all the spindly limbs, and reduce the strong ones to no more than four or five.

The leader or main trunk is also cut back, and the side limbs retained are chosen for location as much as for strength; they should be spaced as evenly around the trunk as possible.

A hypothetical tree with a hypothetical pruning

Good angle

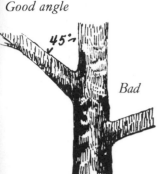

Bad

In later years pruning depends on the tree. Some apples require an annual thinning of suckers and cutting back of too-adventurous limbs. Some stone fruits need scarcely a snip. The object is to keep the interior of the tree as open as possible and to keep the top cut back so that sunlight can get to leaves on the limbs, where the fruit is borne. You want these bearing branches to be as near the ground as practical to make picking easy, so it makes sense to prune out sprouts or limbs that tend to grow straight up, retaining those that grow out in a horizontal direction. One caution, though: the "flatter" a limb grows, the weaker it is. A limb growing at no more than a forty-five degree angle from the trunk is going to be strong enough to hold up a full load of fruit for years. Limbs growing at a larger angle can go in a high wind or under a meeting of the full Cub Scout pack in a tree house. It is a good idea to keep these "flat" limbs pruned in close to the trunk.

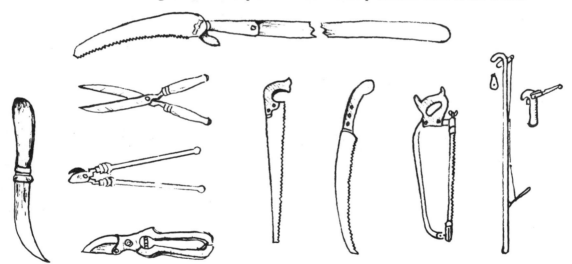

You wouldn't really need all these pruning tools, but having the right tools makes orchard maintenance a whole lot easier.

One solution to a weak limb problem (of any sort) that we've used is a bridge graft. You find two branches, one from the weak limb, another from the nearest strong limb. Twine the two branches together as tight as you can without splitting the bark or cracking the wood. Cut off all leaf twigs but those at the branch tips. Then tie the wrapped limbs well with some organic material that will hold up for a time, but rot off in a year or two (baling twine is good), and coat the joint with tar, grafting wax, or a heavy, nontoxic paint. In time the branches will grow together and the weak limb will have a bit of added support.

Bug Control

No one but a chemical manufacturer would deny that all the spraying of apple and other orchards in this country over the past few decades has completely upset a portion of nature's balance. Now, I haven't quite hit the age of forty and I can remember as a boy going out to my great-uncle Will's farm in Indiana and picking totally unsprayed apples that were wormless and near-perfect. Not so these three decades later. The sprayers have somehow altered the bug populations to the point that natural predators can't keep up with the apple maggots and codling moths and I don't know what all else. You *cannot* grow an apple anymore without some kind of special pest control measures. I'm sure this fact makes the pesticide makers as happy as it makes Louise and me hopping mad. But we may be switching attitudes with them soon. Scientists and organic experimenters around the country are coming up with more and more *natural* pest control measures, and I'm hoping it won't be too long till even the hard-pressed commercial fruit growers can abandon poison sprays.

Air pressure sprayer

The first part of our antibug program, conducted in earliest spring, is to chip and scrape trunks and limbs to remove bark scale that is a favorite egg-laying and wintering-over spot for bugs. Then, contrary to accepted organic cultural advice, we don't mulch fruit trees once they come into bearing. The mulch is just too handy for bugs. I keep the grass cut low and the ground bare as possible in the orchards. Our preferred grass cutters are sheep, but the years we don't buy spring lambs, I use the Gravely tractor and rotary mower to keep the grasses down. This way, the birds can get to ground bugs and those that escape have pretty slim pickings in the plain sod.

CLEANUP

Each tree receives two sprayings with miscible oil—pure, harmless mineral oil that will dissolve in water—one in spring just as the buds are beginning to swell, and the second just after they have cracked open, but before the petals have actually popped out. The second go-round, a spray of wettable sulphur and rotenone, follows the oil. Then when the petals have all fallen, I spray again with the sulphur, rotenone and Ryania, another natural insect repellant. Now again, some may object to even this spraying. But all these things are harmless to humans. Sulphur occurs naturally in underground deposits the same as salt, and Ryania and Rotenone

SPRAYS

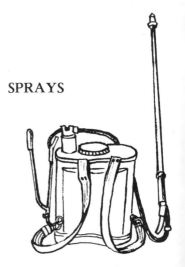

Knapsack sprayer

are natural insecticides derived from tropical plants. We consider them organic enough for us if we can't avoid some kind of spray.

The miscible oil kills overwintering eggs of many bugs, scale in particular. The sulphur keeps down several fungus diseases and combines with the oil to kill off mites that can ruin developing buds. Rotenone and Ryania are effective against just about any bug that chews or sucks on plants, but it doesn't harm most natural predators. We have found that the sprayings keep down aphids long enough for the young fruit nubbins to develop to large enough size that the aphid's bite won't deform and ruin the

One of our allies in natural insect control: A mamma wren on her way into one of the wrenhouses located in the fruit orchard.

fruit. By then the natural wasp parasite and ladybird population has built up enough to keep the little bugs in natural check.

The Ryania is particularly effective against the first generation of another immigrant pest, the codling moth. Since coming over from Europe in the early nineteenth century, this "apple worm" has done its work on just about every fruit that grows on trees. Massive spraying has only sent it farther afield, and it seems to develop immunity to each complex new chemical as soon as it's used. Probably the best control against the adult moths are nesting birds—wrens, starlings, purple martins—that we attract every way we can to gobble up the little gray moths as they fly around on warm spring evenings to lay eggs. The Ryania seems to get to a goodly number of the moth larvae before they can burrow up into young apples, there to start with the core and eat their way out to the place where you invariably take a first bite.

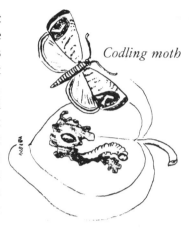

Codling moth

Probably more important than the spray is our twice-weekly trip through the orchards to pick up all premature ground falls the sheep have missed. Most falls are caused by one or another worm. If left on the ground, the worm would burrow out into the soil, pupate, and emerge as a new egg-laying generation. This chore helps against the codling moth and also against the apple maggot, the tiny larva of a kind of fly which is just as happy making its network of thin brown tunnels in stone fruit and cherries, as in its namesake. All the livestock relish the fallen fruit, goats and rabbits as a snack, the hogs as a major part of their spring diet.

GROUNDFALLS

Other Antibug Measures

It seemed futile for us to be cleaning up all of our ground falls while seedling apples and plums were growing wild around us to act as bug nurseries. So, one spring I took the ax and made a tour of our woods, girdling just about everything that sprouted a blossom. This wasn't as criminal as you may think; none of the trees was worth much, being in the final stages of getting crowded out by second-growth hardwoods and firs.

Before settling on the spray and deadfall picking program, we tried using Trichogramma parasites against the codling moth. These tiny wasps lay their eggs in eggs or larvae of just about any moth or butterfly, and the developing wasplet feeds off the host

PARASITES

egg, killing it. I assume they did their job, but the damage from aphids and the apple maggots was so bad, we never had a chance to judge. Still, since we began using Trichogramma I have not seen a single tomato hornworm in the gardens, and the once-abundant supply of cabbage and broccoli worms (loopers) has just about disappeared. Whether the Trichogramma makes its contribution to the fruit orchards or not, their work in the vegetable garden will assure they will continue to have a job with us. I have a little phial of the eggs shipped up from Texas each week from mid-April to mid-September. The tiny bugs hatch in a day or two after arriving and I release them at the upwind end of the garden and orchard area. I won't give the name of any particular Trichogramma breeder, as most so far are pretty small operations and none too efficient, in our experience at least. Still, they have the right idea, and ought to be patronized by organic enthusiasts, even if the price of a phial of bug eggs is a whopping one dollar per—and that's in quantity lots.

Trichogramma wasp

There are several other steps we take against other specific bugs. The curculios, apple and plum types (the latter bothers all stone fruits), look like boll weevils: sort of round, gray or brown little bugs with oversized, curved snouts. To control them we make extra sure to pick up deadfalls in the month of June—the larvae are buried inside—and to knock off any prematurely coloring fruit. These will often mummify, staying on the tree with a wee curculio inside.

BORERS

The several species of borers are highly susceptible to the Trichogramma released in summer and early fall. They tend to be worst in dry years and on sun-scalded young trunks. I keep young stock well watered, as mentioned earlier, and trunks are kept wrapped with foil to prevent sun damage. Where a rare gummy spot or hole with a little mound of sawdust appears on limbs or at trunk bases of an apple, peach, or other stone fruit, I chip away the bark till I know which way the burrow is heading, then I heat up a length of coat hanger with the propane torch and jam it in as deep as it will go. If the hot wire doesn't incinerate the grub, the steam from boiling sap sears it. The hot wire seals up the wound in the bark too.

Earwig

As of this writing we are in the midst of a great earwig population boom. These little brown crawlers with pincers on their rears are found under rocks most of the time, but lately they have appeared in huge numbers—due, I suspect, to the unusually cool, wet springs we've had during the early 1970s. Earwigs are great fanciers of dead-ripe fruit, especially our peaches and pears. When the fruit are beginning to flesh out well, I just apply a thick ring of tanglefoot on a length of building paper around the base of each tree. If the smelly, gummy stuff is renewed every so often, it repels or catches the earwigs and a good many other unwelcome bugs trying to climb the trunks.

Our latest organic pest control measure is another bug   BT
disease, *Bacillus thuringiensis* or BT, being sold now to home gardeners as a bottled solution under the trade name *Thuricide*. This disease attacks the innards of leaf-chewing caterpillars such as tent caterpillars and the horrendous gypsy moth that has been known to defoliate whole forests. We spray it on any tree that has a colony of tent caterpillars, more of the inch-worm-type leaf eaters than normal, or just about any oversupply of caterpillars of any sort. It is nearly 100 percent effective within a week's time.

I plan to spray BT on all the fruit trees early next spring, perhaps combining it with the rotenone, Ryania, and sulphur schedule. Sooner or later we'll hit on a combination that turns out a crop of all-perfect, supermarket-style apples and pears. When that happens, we'll know we've gone too far and will begin cutting down on pest control measures. You can be sure the first to go will be the chemicals, comparatively good chemicals though they may be.

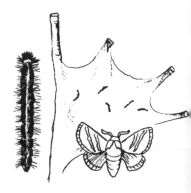

Tent caterpillar

7

The Outdoors After the Bugs

Through most of May and often into June, such garden chores as thinning carrots and hoeing the peas are restricted to the hour or so just after dawn when the biting black flies are sleeping in. When the sun comes up so do the bugs. But then there comes a spell of warm weather that brings out the mosquitoes to make dawn and dusk miserable, but it also ends the black flies' sojourn for that year, and the outdoors belongs to us again.

First I empty the cold frame of early-started plants. Tomatoes are first, often coming from the cold frame two feet high and well blossomed. I dig a shallow trench for each plant and lay the peat pot and most of the growth in, covering with perhaps two inches of warm surface soil. All that remains above ground (under the hotcap) will be six inches of plant crown. The reasons for deep planting are two. First, the buried stem will develop loads more roots for food and water gathering. And the fruit that comes on the first set will be borne low down, not up high on a spindly, early-spring-grown stem.

TOMATO VINES   We've tried all sorts of methods for supporting tomato vines so as to keep fruit away from the ground where it can be affected by soil-borne rots and half-eaten by the mice that love to burrow through our deep late-summer mulches. Staking on poles is never fully satisfactory as our organic soils produce tomato trees that quickly outgrow a single pole. One year I tried supporting the early varieties with a dwarf determinate vining habit on old cot springs scavenged from one of the old eighteenth and nineteenth century dumps behind our stone walls. The vines stayed above the spring lattice but the fruit hung below, so getting it took a lot more groping around than if I'd just let the plants grow as they might.

Now we've settled on wire fence cylinders to support tomatoes. Standard lightweight farm fencing, five feet high with wires spaced in two-by-four-inch rectangles, is cut in six-foot lengths that are formed into cylinders. Once the tomato plants have put on a couple of feet of new growth I set the fence supports over them, and tie the top of each plant to the wire. Then as the top leader grows and as suckers branch out from the axials of the leaves, they are tied on around inside the cylinders. In time the whole wire enclosure will fill with stem and leaves, but the clusters of flowers, followed by fruit, pretty much hang outside the supporting wire. There they receive plenty of sun for quick ripening and are easily picked.

The early varieties with determinate vines are supposedly unsuited to staking, as maximum production is obtained by letting them branch out as much as they are able. The wire enclosures allow this branching to take place, but the fruits are kept off the soil. The wire supports also make it easier to prune the indeterminate, rank growing, mid- and late-season varieties. Along about mid-August I go through the tomato patch snipping off all terminal sprouts and any suckers that have not set fruit. Thus the plants can put all their energy into maturing fruit before frost.

Tomato hornworm

Outside of mice and chipmunks that do get to the lowest fruit on occasion, we have very few pests of tomatoes. Annual relocation of the plots prevents the several rots and blights that commercial gardeners and farmers must worry about. Since our soil and plants are poison-free, natural bug controls have free play. Honestly, the only tomato hornworm I have ever seen was a huge, green, crawling corpse, its back covered with the white eggs of one or another wasp predator just about to hatch and consume their obliging host. I put it in a jar to watch the process. Inside two weeks it was an eaten-out hulk and the garden was hunting ground for perhaps a hundred more skinny-waisted wasps.

PEST CONTROL

A common nuisance of young plants of many kinds are the cutworms. These larvae of night-flying moths hide out in the soil or sod during the day, then work just under the soil or above the surface, often girdling plant stems as they eat. If we had a serious cutworm problem I'd just plant young seedlings and transplants in tin cans with both ends cut out. But we have almost no problem, though a good many neighbors have lost entire plantings to cutworms. Of course I may live to eat my

Cutworm

words—never can tell about nature—but to date we lose only one or two plants a season to obvious cutworm damage. I suspect that some credit must go to the frequent tillings our soil receives, which expose the grubs to hungry birds. Another factor must be the moles; our gardens are open at any time to the furry little burrowers that live on grubs in soil and sod. And the above-surface grubs that do survive the moles are exposed to our toads—dozens of them that hide under the mulches or hot tents during the day, to come out at night and spear any bug that moves with their sticky tongues.

And isn't this the essence of organic agriculture? The bugs get their share, but with most bugs and most plants their toll is minor, any really serious damage being prevented by perfectly natural controls.

Cabbage, Broccoli, and Spinach Combination

By late spring the early plantings of spinach have been thinned to go into salads and cooked greens. The plants, now growing rapidly, stand about six inches apart in the row, and are just a few weeks from harvest. Right into the spinach planting I set the **Green Comet** early broccoli and the early **Jersey Wakefield** cabbage, rows and plants in the row both two feet apart. The slow-growing members of the cabbage family will not compete with the spinach, but will be well-established when the greens are harvested. And the pairing tells us when to pull the spinach; when the dark-green crumpled leaves begin to obscure the blue-green, foot-high cabbage leaves, the spinach is getting ready to bolt to seed and must come out.

SPINACH
STRIPPING

The spinach hits its prime usually the second-to-last week of June, typically the middle of the first steamy and humid heat wave of the year that brings out both the mosquitoes and the worst in the homesteader's disposition. To get the harvesting over as quickly as possible I long ago devised the technique of "spinach stripping." The plant is pulled, I grasp the stem and several withered or bleached lower leaves with my left hand. Then with the right hand I just strip off remaining leaves and tender stems. It takes a bit of practice to develop the correct pressure, but after a

Gardeners don't always smile. The spinach—or other crop—has to be picked when it is ready, regardless of heat, humidity, and pesty insects. Stripping the spinach, as I am doing here, speeds the job.

few false starts I'm getting the good leaves and tender stem parts into the gathering cart while stalk, roots, and tough stems go into a pile for the livestock's evening meal.

Unless the spring has been unusually wet and overcast, four LR Squares will give us about thirty pounds of spinach. Blanched in boiling water for a couple of minutes, chilled an equal time, then packed in pint freezer containers, that gives us one or two meal servings a week for the months between frosts. Now, you might be tempted to think that we've put a lot of time and effort for not much in return, so let's try putting a dollar value on it all. At this writing, fresh supermarket spinach (Clordane, Fermate, Paris green and all) is selling for forty-nine cents a pound and rising. So our thirty pounds is worth nearly fifteen dollars retail. Take out a dollar for cost of seed and equipment to get fourteen dollars and divide by time spent—say, a half-hour in ground preparation and planting, the same in cultivating, plus an hour in the harvest and freezing. That's two hours at fourteen dollars, or seven bucks an hour, which isn't so bad considering I can name my own working hours, don't have much of a commute, don't have to pay taxes on the money saved, can tell my supervisor where to get off any time—and besides, I love the work.

YIELD

You might suggest we also deduct the value of cost of freezers, containers and all, but I'd consider it a wash against time and gas money *not spent* shopping for equivalent items. One thing we can add in, come to think of it, is the cash value of the recreational benefit of the gardening experience. The hours we spend bending and stooping and sweating over the land, your average suburbanite spends golfing or swimming at some expensive club. But while we are creating something of cash value with our exercise, he is spending money to have his. Now, I do not begrudge these good folks their country clubs. If the only way to have a good time and get exercise is to pay for it, that's life in our times. Louise and I used to live that way too, and even now she's not at all loath to unlimber her tennis backhand on a Sunday afternoon. But in the days we had to buy all our exercise I figured the cost at about three dollars an hour, all things considered. Add that cash saved to the money not spent for spinach and my ''wages'' come to a good ten dollars an hour. And into the bargain, our spinach is fresh, a home-garden strain with superior flavor and vitamin content, and it's grown with not a trace of potentially harmful chemicals. And that latter quality is nearly priceless these days, don't you agree?

The Organic Method

Avoiding harmful chemicals is just one part of organic food production; the other, equally important, part is using only natural plant foods. Which brings me to the second part of the lecture, which I'll try to make brief. Why use manures and plant residues and natural mineral sources that weigh tons and assay out to only a few percent nutrients, when chemical fertilizers are of high assay, low weight, and come in easy-to-use sacks? Scientists insist that the nitrogen, potassium, etc., that plants need are absorbed in pure chemical form. And that it makes absolutely no difference whether the source is natural or artificially made. Well, I go along with that.

HEALTHY SOIL   However, the reason isn't that the three basic plant food needs are chemically different in an organic loam. It's that a natural soil, teeming with microorganisms and worms and plant fiber, provides a complete plant diet, the trace elements that we know all life needs, and probably a lot of other food components science hasn't discovered as yet.

There are other advantages to a plant of having its roots deep in an organic soil. The burrowing creatures aerate the soil, permitting better circulation of both air and water. The organic grower adds mulches in hot weather, keeping plant roots cool and conserving moisture. He or she also continually contributes garbage, old leaves, and animal manures either as is or as compost and ground rock for minerals. All of these disintegrate into basic plant foods slowly but continually so that the plant is able to absorb just the amount of each food element it needs at each stage of its life. Often the more concentrated chemical fertilizers will cause abnormal growing behavior. Adding too much chemical nitrogen, for example, will cause tomatoes to grow great and luxuriant stems and leaves, but fruit production is greatly reduced. In an organic soil, or in ours at least, the plants also grow luxuriantly, but the fruit production is also phenomenal.

So much for the advantages to plants. Growing with natural fertilizers is of increasing importance to mankind. Know what the most important component of chemical fertilizer is? Petroleum, both as a raw material and as the industrial power source. We needn't go into the shortage and high cost of oil again. Environmental pollution is greatly reduced with the use of natural fertilizers. For one thing, it is a way of recycling material that would otherwise end up a prime source of water pollution.

THE ENERGY SIDE

So we can conclude the lecture by stating that using organic fertilizers is a positive contribution to reestablishing the environmental balance in a world of too many people, thus too few resources and too much pollution. And growing food that is superior in vitamins and minerals and completely free of unnatural chemicals is about the best way I know to assure that our children will grow up strong and healthy.

Feeding Baby

Speaking of feeding children, I know of at least two full-length books on why and how to prepare and feed babies healthy, home-grown foods. Perhaps there is more to it than we are aware of, but all Louise could come up with was a page or two to describe how we are feeding a pair that enjoy howling good health on home-produced organic foods and pure spring water. It's a fact that when

OUR HEALTHY KIDS

there is a bug around and most other kids are flat on their backs, ours maybe have a sniffle or two and tend to spend the nights trading awake periods. Last year the family had a case of round-robin Asian flu which put us out of commission for almost a month. The problem wasn't the children, who barely showed the symptoms, but the older generation. The only medication they receive regularly is a tooth-strengthening fluoride. And though they get regular check-ups, each has seen a doctor for just one minor problem apiece, and that mainly to reassure Louise and me that we were doing the right thing. No, I'll have to take that back. When he was two, Sam got loose in the garden and consumed a whole hot, red pepper. It took a while for the heat to get through to him, but when it did he let us know it. Louise called the doctor who suggested she wash his mouth out, then complete a good Italian meal with sausages and mozzarella cheese. He said it might sting, but the pepper was good for his health.

THEIR DIETS

The doctors also endorse another of our kids' staples, goat milk and honey. Infants who don't do well on condensed milk and sugar formulas are most often put on what our babies had from the beginning. Goat milk and honey are more easily digested than cow's milk and refined sugar, and what better recommendation could it have than Dr. Spock?

About the only regularly purchased foods we feed the children have been baby cereal and orange juice and an occasional supply of fruits or vegetables we did not have home-raised, such as sweet potatoes. But the vast majority of the children's diet is made up of the same good foods we eat ourselves. The first few months of their lives, their food was pureed in the Oster blender. Then when they were about six months old we begain making junior food with the Foley food mill.

Feeding Martha.

All the fruit except strawberries and green vegetables of all kinds were early favorites. Even broccoli, though most baby books claim that little kids don't take to the strong flavor. Maybe Dr. Spock never tried his patients on organically grown broccoli. Our good potatoes, blended up skins and all, winter squash, beets, and carrots come from the cold cellar to be gobbled up. Louise cooks up a batch of one fruit or vegetable, blends or mills it, and either refrigerates or freezes any surplus in one-meal portions in an ice tray.

DRIED FRUITS

One of the kids' favorite foods, and ours too for the help they provide in teething, are fruits that we home dry. Apple rings, apricot halves, pear and peach slices, are all dried down to a good leathery state and passed out to the teething machines as each new gum bump appears. At first we dried the fruit in the sun, just placing it between two old window screens and turning it frequently. But the dried fruit became such a staple that we got a real food dryer. Plans, a book called *Dry it You'll Like It*, or a dryer kit are available from fellow organic homesteaders Gen and Bob MacManiman of Fall City in Washington State.

Young Sam makes a hit-and-skeedaddle raid as Louise unpacks apple rings and apricot halves from the food dryer.

**USE THE
WHOLE FRUIT**

The dryer lets us use parts of fruit that would otherwise go to the pigs: cores, skins, and even stems. Louise saves a batch of leavings, say the cores and ends from a half-bushel of Macoun apples we have cut into rings. They are steamed for a few minutes, blended up in small batches, then mixed with spices: apple pie spices for apple leftovers, or a little lemon juice and honey for pear cores, for example. The paste is poured out on waxed paper and dried. The resulting fruit leathers, as they are called, can be cut into pieces and used as a hiker's quick energy source, a nutritious sweet treat for kids (we pass out fruit leathers at Halloween), and for Christmas candy, gifts, and just for good chewing fun.

**HEALTHFUL
SWEETS**

All children like sweet things and most parents maintain a running battle against Hostess Twinkies, breakfast cereals that are half candy, and other tooth-rotting packages of empty calories. On the homestead there is no such problem. Our sweet breads—gingerbread, apple crisp, pumpkin bread, and the like—are all made with whole grains, protein rich seeds and nut meats, and natural sweeteners like honey, molasses, and fruit. The children can have all they want. Our version of a hot-weather favorite, popsicles, is also packed with good nutrients, not a mixture of water, sugar, and chemicals as is the commercial variety. We mix up fruit juice, complete with pulp, add in enough honey to give it a good frozen consistency, and freeze in ice trays or paper cups. A stick goes in the middle. Again, the children can live on the things if they feel like it. They would eventually tire of it, so long as it is not a forbidden luxury, and place it in proper perspective just as any other food.

A Child-Raising Ethic

Louise and I feel that this attitude toward sweets, on the part of both parents and kids, points up what is best about raising children on an organic homestead. Sure, we plug up electrical outlets and fence crawlers off from the wood stoves. But basically, the children are free to be themselves, to act completely naturally in a completely natural setting. Which is to say they will yell and cry when they feel like it and likely turn the house to sawdust before they get into grade school. But hopefully they will grow up with a conserving and loving attitude toward the world they share with other creatures, and an objectively critical outlook on all the frummery that is touted as the "good life" to much of society.

At the risk of sounding too self-satisfied, I suggest that this wholesome, natural attitude extends to all of life for a homestead child. The family works and plays together. Mom isn't out at a bridge club or whatever to give her something besides house work to occupy her mind. Louise is making pots, sewing, or putting up food in spite of the help provided by toddlers. And Daddy doesn't have to desert the family every day. I am in my study or in the fields, often as not with a small helper to make the job twice as long and ten times more fun.

Daddy's little helper.

We can't help but think that a child who at age two was helping gather grass for goats, at three was gathering his own breakfast egg, who has participated in all the most basic chores of living, will be served in later life no matter what his choice or career or way of life. He will have grown up directly involved with nature with a first-hand knowledge of life lived at an elemental level. He will have his feet in the real world, partner with the land from the beginning.

Turning Sod to Organic Loam

About the only homesteading activity that the children never even get close to is work with the power equipment. The shredder, Gravely tractor, or the tillers make enough noise that an approaching child could get a hand in tines or flails before an operator knew it. So most land preparation is a solo job.

When we first moved on to the homestead both our garden and field locations were grown up in a thick sod of perennial meadow grasses. If you cut out a hunk of sod and turn it over, in two days the grass will have sprouted up and in two weeks you can't tell the earth had been disturbed at all. So our battle against this stuff—quack grass, Johnson grass, or witch grass are several common names—is never ending.

BREAKING SOD

Our first piece of equipment was the Gravely tractor mentioned earlier. In addition to the rotary mower we obtained a rotary plow, an arrangement of four steel blades that whirl rapidly, chopping up just about everything that gets in their way up to a foot deep. What it can't chop it throws out, including rocks the size of your head. Needless to say, it is a hairy monster to operate, though it does a fine job of initial soil preparation. The next tool was a small tiller, with tines in the front and a drag stake in back to keep the tines working. It does a fine job of shallow cultivation; I have the outer tines on each side removed so it only tills about a foot-and-a-half wide path.

The most recent addition is a big tiller with automotive-type gears and the tines in the rear. It also has a snow plow and a furrowing attachment. Several companies are making them now as gardening grows in popularity. Ours is a Troy-Bilt by the same

Ancient history by now, but here I am our first year on the place, churning up sod with the Gravely and rotary plow.

company that made the original Roto-Tillers, Garden Way Manufacturing of Troy, New York. If we were restricted to only one gardening machine, the tiller is the one I'd choose. It doesn't go as deep as the Gravely, can't till as narrow a strip as the other tiller, but it pulverizes the soil better and is many times faster and easier to operate than either.

A good tilling spring and fall, inter-row cultivation several times a season, plenty of mulch and manures have turned our gardens to a rich, deep loam that is the organic gardener's pride. The soil had lain fallow for years, of course, and so was rich to start with, but it gets better with every passing year. There are two main gardens, about sixty by thirty feet in size on flat areas part way up the hillside. Above and around them on the slope itself are several smaller plots, about fifteen feet long and six to ten feet wide. We leave strips of sod between these patches to keep the sloping soil from shifting down hill. Without the soil strips we would literally have to truck the bottom part of the garden back up to the top every few years.

The two main gardens receive the most compost and care and in them we grow the most demanding crops. Each year I let one or two of the strip gardens go back to sod to lie fallow for a year or two, and cut out a new plot or two. Into the new plots will go the hardier, tall growing crops such as corn, pole beans, and peas. The grasses are kept honest by frequent inter-row tilling and old hay mulches till the food plants are tall enough to shade their own feet, keeping weed growth down.

Every few years I test the soil of each plot just to see if nutrients are in balance and acidity is in the desired range. The tests are made with a Sudbury Soil Test Kit, available at most hardware or feed stores, by mail from seed sellers, or direct from the maker, Sudbury Laboratories, Sudbury, Massachusetts. The kit tests for the three major plant nutrients, nitrogen, potassium, and potash, as well as Ph or acidity/alkalinity. Sudbury sells chemical fertilizers in a highly concentrated form, but each kit also contains an organic supplement prepared in collaboration with the editors of OGF.

If I find any glaring inadequacies in a plot they get corrected immediately. Our soil is naturally acid, so it must be sweetened frequently by an application of ground limestone. We use a variety of limestone called dolomite. It contains a good supply of magnesium, an important trace element. With the manures and cover crops we use, I seldom need to increase

REGULAR SOIL CARE

Working with the Troy-bilt.

SOIL TESTS

nitrogen anywhere but in the corn field. I'll get into that shortly. To increase phosphorus I scatter on a dusting of steamed bone meal before the spring tilling. Almost a quarter of its weight is available phosphorus. Wood ashes from the stoves provide all the potash we need—8 percent of the ash is the essential plant nutrient.

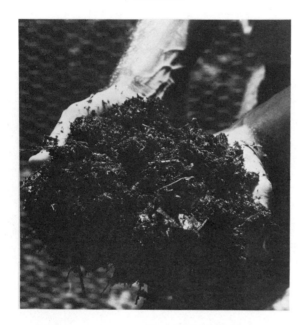

The secret of every successful organic gardener: Compost. This sample is former chicken litter, after a year under cover.

Warm Weather Planting

Once the weather is warm enough that heat-loving seed will germinate, I spend a busy week getting in the postbug plantings.

Late Potatoes

Late potatoes are the old-time favorite **Green Mountain**. Like the early variety, **White Cobbler**, this spud has just about been replaced by the new plastic varieties. Its tubers come in all sizes and shapes, and it doesn't keep really well. It just tastes like a potato. I put in fifty pounds each year and harvest three to four times that, sometimes more. The procedure is the same as with the early variety except that a thick hay mulch goes on right after the tubers go into the trench.

Beans

Beans are second only to peas and sweet corn as a vegetable that best shows the difference between home-grown and store-bought. We grow a couple of rows of **Eastern Butterwax**, a lovely tasting yellow variety, about the same amount of the flat-podded green bean, **Romano**, and the Stokes special medium-flat green variety **Green Crop**. But our main green bean planting is **Kentucky Wonder** white-seeded pole beans. The flavor can't be beat, crops are huge, and when we have all we can eat fresh, freeze or give away, we let the pods mature and shell out our winter's supply of dry baking beans. The bush beans are planted in rows two feet apart, seeds thinned to four inches apart. For the pole beans I put up teepees of four seven- or eight-foot aspen poles tied together at the top with baling twine, the poles set about a yard apart and pushed into the soil. A half-dozen seeds are planted around each pole, thinned to the three best when all are up. It is difficult to hoe around the poles, but some sort of weed control is needed, especially when the beans go into a newly plowed section of garden. Once the soil is good and warm, I lay down several thicknesses of newspaper and cover it with old hay, mainly for appearance's sake. The beans love it.

Beans, just starting up their poles.

Sweet Corn

Going alphabetically for a change, we've already covered beets, the members of the cabbage family that we grow, and carrots; so at last we come to sweet corn. This is the one vegetable that should be picked as near to dinner time as possible to retain the fast-fading sweetness and fresh corn flavor. We literally have the pot boiling. Then Sam and I go up to the corn plots and pull up whole corn stalks. On the way back to the house the sugary stalks go to a most appreciative pen of goats. We husk the ears, giving the tiller leaves to the rabbits, the silks to the chickens. Then after

Hustling with the corn, the boiling pot awaits.

Corn earworm

we've feasted on the kernels, the buttery cobs go out to the pigs. Nothing's wasted.

Till recently we thought we had a private corn secret. We grow only hybrid white corn, Joseph Harris's unexcelled ninety-four day variety **Silver Queen**, and this year we'll try their new introduction **Spring White** which comes on in a little over two months from planting. I say our secret is out because the folks at Harris have finally admitted that the whites are the best corns they offer— ''probably the best we have to eat,'' they say in the '74 catalog. We've tried the bi-color varieties and the super sweet ones and plenty of others, but none match the tenderness and flavor of the white corns, in our opinion.

Hybrid corns are comparatively delicate plants. Each variety must receive pollen from one of its fellows or the ears will not develop as expected. We plant corn in blocs in several of the small strip gardens, rows a good three feet apart, put seed a half-inch into the soil, and when the sprouts are up we thin the small-stalked early variety to a foot apart, the later one to a foot and a half between plants.

Since we never let corn stalks stand over winter, we have no problems with corn borers or earworms that spend the cold months in old stalks. Crows delight in pulling up newly sprouted seed, though, and to stop them I tie strips of foil to sticks that are pushed into the soil at a sharp angle. The slightest puff of wind, even from an approaching bird's wings, makes the foil sparkle enough to keep the crows at a healthy distance.

Cucumbers

Cucumbers are planted in the same type of wire fence cylinders we use for the tomatoes. They will grow naturally up the fence. A half-dozen seeds are planted around each circle of fence, and we thin to the three or four best plants. Variety choice has been a problem for us, and there have been years that the pickle supply was pretty skimpy. The problems are the several mildews that somehow came to join us a few years ago. The whitish powder covers the leaves and the plants wither away.

Fortunately, the plant breeders are coming out with varieties that are naturally resistant to most cucumber diseases. This year we will be trying two such varieties newly offered by **Stokes**. **Sweet Slice** is a good new burpless variety developed originally in Japan. For the pickle jars it will be **Pioneer**. Both varieties mature in under two months and are gynoecious, meaning that all flowers are female, and the yields are huge and long lasting if the vines are kept picked. The only disadvantage is their lack of male pollen, so the seed companies include a small percentage of other male-flower producing varieties. These seeds are dyed so the grower will know to plant one every so often in the row. I mark the male seeds with a stick, so as to be sure not to thin them all out.

The Herbs

The herbs are Louise's specialty; it takes a special patience to put up with the tiny seeds, long germination times, and slow growth of many. She starts them inside in egg carton flats, watering and tending them like children till they are big enough to go out to her herb garden under the rear kitchen windows. Sad to say, with two little garden helpers the herb garden has suffered a bit of late. But her perennial sage, tarragon, lavender, and chives manage to keep going. The dill is a big plant and enough survived last year for the pickles. But it may take a few summers before we can look forward to basil, summer savory, or sweet marjoram.

Okra

Okra, the essential ingredient in the Creole dishes that are one of my own cooking specialties, is thought of as primarily a crop suited to the deep South. But we do very well with it here. The big seeds are planted in rows three feet apart, plants thinned to stand a foot apart. In two month's time lovely purple flowers will appear, followed quickly by pods that will grow woody in short order, and so must be watched closely. We use the earliest variety offered, currently Stokes's **Perkins Mammoth Long Pod**.

Hot Peppers

Hot peppers we don't grow just now since Sam got into one. It's a matter of personal taste, but we use perhaps a half-dozen bell peppers a year. They freeze well and keep forever and we still have an ample frozen supply from years back. When we come near to running out we'll look over the varieties offered, and probably select the most disease-resistant.

Pumpkins

Pumpkins are planted along the south end of the corn field, seeds of the small variety **Spooky** put in every yard or so. The fruit, of course, is essential at Halloween. Louise regularly bakes a half dozen, freezes the pulp for pies, and we roast the seeds. We tried putting pumpkins in the sweet corn, but found we couldn't pull the stalks for the animals without harming the pumpkin vines. If pumpkins are prone to any diseases or insect pests, they've never come our way.

Radishes

Radishes go into so many rows of slow-germinating plants early in the season that we don't make special plantings till the weather cools down in fall. I really have never found an appreciable difference between varieties, so we order the fastest-maturing variety of red radish we can find. White radishes we've never had much success with but will try if I ever get around to putting in a rock-free, raised planting bed. We do lose a lot of radishes to root maggots, so I just plant a double supply, seeds about a half-inch deep in rows a foot apart.

Turnips

Turnips aren't a big favorite with us, though I do occasionally put in a few rows of summer turnips if we are having an unusually wet fall. **Purple Top White Globe** is the variety, rows a foot apart, the tremendously vigorous seedlings thinned to stand about four inches apart. The tops usually get pretty well chewed up by flea beetles that late in the season, but the globes have a fine, mildly spicy turnip tang when boiled and served with butter, salt, and pepper. Louise also needs them for her Canadian pea soup. Two or three carrots and the same number of turnips are boiled up with several cups of dried sweet peas and some strips of salt pork. Wonderful stuff for lunch on a cold winter day.

Squash

Squash is another big favorite with us, especially the crisp and wonderfully mild summer variety zucchini, whichever of the many reselected strains of it sound best each year. We've found winter squash to be a notorious cross-pollinator, so have quit trying to plant more than one variety. Till recently we have alternated between the turban-type **Buttercup** and the acorn variety **Table Queen**. These are rank-growing vine plants that take a great deal of room. But there is now a pair of bush varieties, Harris's **Emerald** of the buttercup varieth and Stokes's **Table King** bush acorn. I hope they have the quality and keeping ability of the vining kinds. If not, we'll be switching back.

Field Crops

Planting large concentrations of one or two grain crops for the livestock and maintaining them under organic growing conditions is asking for trouble both from insect infestations and soil depletion. So far we have managed to avoid both problems by practicing a controlled version of the old-time way of using a field for two years and letting it lie fallow every third year.

OUR ROTATION

Our fields are divided into thirds. The first year one-third is planted to fairly sparse crops of grain—to date only field corn and sunflowers. The second year it is planted so the green manure crop buckwheat. Buckwheat is a strong, vigorous grower and keeps weeds from coming back. The third year each section is planted to alfalfa which is a legume and fixes nitrogen in the soil. So each year we grow corn, buckwheat, and alfalfa on rotating sections of field. The only crop thoroughly harvested is the corn. Years we have a long season between frosts the buckwheat may mature enough for us to hand harvest enough of the brown, three-cornered, hat-shaped seeds to be roasted, then ground up to add their earthy flavor to hotcakes and breads. But most years the only direct dividend we get is when the bees gang up on the flower heads and award us with a super of dark honey with the special smoky buckwheat flavor. As for the alfalfa, I usually scythe a batch every morning after the wild grasses around the place begin to thin out. But mainly both the buckwheat and alfalfa are to add nitrogen and other nutrients to the soil. Each fall after the corn is harvested, I go over the whole field with the rotary mower—twice over the corn. Then I either till in the residue with the Troy-Bilt or ask a neighbor with a tractor who does a lot of garden plowing in the area to run over the land with his disc harrow on his next trip past.

The disking is repeated as early in the spring as the ground warms and dries out. Then the seed goes in, alfalfa and buckwheat broadcast with the same rotary spreader I use to apply lime, the corn in a small one-drill planter by Planet Junior that I

One-drill planter

have rigged up to attach to the front of the tiller. The corn rows are spaced a bit more than a tiller width apart so cultivations will be easier. The tiller, set shallow and at its highest speed, is run over the other two sections to get most of the seed into the ground. Then a set of anticrow strips is put in. The big tiller stays in the field under a tarp and is run between the corn rows each week till it is hip-high and shading the soil.

VARIETIES   Selecting a variety of buckwheat isn't hard. There is only one. Any feed store can supply an alfalfa seed (or other cover crop) suitable to any good growing area in the country. But choosing corn seed is a bit more complicated. First problem is the southern corn blight that has recently decimated many of the field corn stands in the country. Plant hybridizers have developed resistant varieties, and anyone growing field corn, even in coldest New England, should be sure he gets a resistant variety.

Most field corns are hybrids, sold only in fifty-pound lots at about a dollar a pound. The seed will stay good for years, though. And hybrid corn's maturity is best calculated not by estimated days from planting to maturity (seventy to over 125 depending on variety), but on a complicated system called degree days, or heat units. Basically, so long as the soil and air temperature exceeds 50° corn will grow, and to get the heat units for each day you add the highest day temperature to the lowest night temperature, divide in half, and subtract the 50° base figure. So if the high was 85°, the low 65°, you add to get 150, divide in half to get 75, subtract 50, and your heat value for that day is 25°. Corns come in heat unit requirements ranging from about 1500 to 3000; in other words, some require twice the heat to mature than others. The one selected depends on where you live. Needless to say, we go for the short season varieties.

SILAGE

Another consideration is end use of the corn. Most is raised for silage, cut and chopped while still green and juicy, then fermented for animal feed. We make silage only when the summer is too cold to mature and dry the corn suitably. So we opt for varieties bred mainly to produce grain but not grow high as an elephant's eye. Other considerations include standability, i.e., ability to stand up in rough weather, drought tolerance, which is important, and ability to grow well under crowded conditions, which is not a factor in our soil-conserving planting method.

PROBLEMS

Insect and disease problems are minimal, or have been so far. Alfalfa is heir to several problems but I suspect they crop up only when the plants are overdosed with chemical fertilizers to produce two or more cuttings a year. Corn borers, which dig down into the heart of new plants, have not bothered us so far. If they ever do I will have to spray with the same miscible oil used on the fruit trees. Aimed down into the young corn, the oil smothers the bug. Earworms are no problem in field corn. The animals don't care about the appearance of their feed and one or two worms don't eat much. Smut, where an ear blooms with a black mold, comes now and again. I just remove the whole plant and feed it to a goat. The smutty ear goes to Horrible Pig.

Corn borer

About the only other problem we have with the field crops is the hours of hard work they entail in planting, cultivating, and harvesting. Our first priority is a small farm tractor to do that work in a more reasonable time. I know our forebears did it all with horses or an ox team. But when faced with days of one-row-at-a-time work on acre-sized plots I find the prospect of a cooling drive to the feed store awfully tempting.

Rewards Begin Coming In

And then just as suddenly as it began, the hectic spring and early summer activity is over and the rewards of our labors begin coming in. First are the asparagus spears, then the strawberries. After apple blossom time I remove a honey super even if it is not completely full just to get the spicy apple honey. And then as we get deeper into July and the heart of summer, harvest begins in earnest: peas, lettuce, then the first tomato, often by the next-to-last week of July. And until frost we become virtual vegetarians, living off the fresh bounty of the gardens. Which is a good time to deal with the large livestock.

Completing the Homestead: Large Stock   8

As July rounds into August the trees and meadow plants begin to lose their spring-green brightness and take on the more somber shades of the season of maturing. Apples are beginning to make color, the hickory nuts are swelling, and we find the garden's productivity has us at work most evenings canning or freezing a harvest that grows larger each day. This is the time of year when planting is over, weeds have lost their early-season vigor so the gardens need little cultivating, but there are still several weeks to go before the full harvest is on us. Now there is a bit of time the fields don't have a claim on. Some of it is spent sitting in the warm-summer sun, just listening to the corn grow. But most is devoted to the large livestock. August and early September is birthing time on our place, the animals too producing their annual offspring in keeping with the season's cycle.

We think that taking on large stock is the one step that makes a homesteader a homesteader, rather than a serious gardener with a few chickens in the back yard. But it also is the one step I would caution anyone to ponder most carefully. (We didn't ponder quite long enough one year, as I'll discuss later.) Poultry and rabbits are one thing; their quarters are relatively quick and easy to build, feed and water can be provided more or less automatically, clean-up is simple, and the loss of one or two animals doesn't amount to much.

But care of large stock is much more demanding on both **DEMANDS** time and resources. Feed and water is needed in much greater quantity—hay runs to the ton, not the fifty-pound bag. Clean-up runs to the ton too—of good, organic fertilizer, true, but it takes a whole lot of pitching. And then a producing milch cow represents at least a $500 investment, a good dairy goat $100, a pig $25 and up, and horses can run into thousands. So, the illness of a large animal can amount to a major money problem. And what do you do with the carcass of a steer, say, that just lies down and dies from eating a half-inch length of old fence wire? There you are one fine summer day with a half-ton of meat that is unfit for human consumption lying in the pasture (or worse, in a barn stall), and the temperature is rising into the high eighties. Think about it.

ECONOMICS   Let's examine the true economics of animal raising. Take hogs. Say you buy a weaned pig at two months of age, raise it the added six months needed to reach slaughter weight of about 200 pounds, giving it all-bought feed, and then pay for custom slaughter and cutting. Here's a rough cost estimate as of this writing, and I'll bet the prices have gone up plenty since:

Pig:	$20.00
½-ton feed:	80.00
Slaughter:	40.00

That is a cash outlay of some $140, which means you are paying $1.40 per pound for the 50 percent or so of live weight that comes out in good meat and bacon. You'll also have up to half a hundredweight of lard and pig skin, for what good that'll do you. Even at today's skyrocketing pork prices you would be money ahead to keep buying your ham from the butcher. Of course, if you breed your own pigs, raise some or all of the feed, and do your own slaughtering, cutting, and smoking, it's a different story. But that all takes effort and skill that no one can acquire overnight.

Or how about a goat? Don't they more or less turn grass into milk? Yes and no to that question. For one thing, you must feed a doe kid for a year till she is old enough to breed, then feed her for the better part of another year till you are getting milk. At the rate each of our goats eats—about a quart of grain, odd greens, and at least six pounds of hay a day—you have a hundred-dollar value or more sunk into the animal in feed alone before the first squirt of milk hits the pail. So if you are lucky and the goat produces a couple of quarts a day over ten-month periods for three years, you have the equivalent of a dime a gallon invested in that milk before you ever see it. So, as I say, large stock shouldn't be taken on without a good deal of thought and planning unless you have time and money to burn.

OUR MISTAKES   But once the decision is made, please take it slowly. In our experience it is a mistake to take on more than one new species a year, that is, if you are as raw novices as we were. I say this because we started with hogs and goats the first year after we'd paid off our mortgage and were able to devote only part time to money-making activity. I planned to run the pigs and goats together in an electric-fenced pasture till time and cash could be found to build a proper goat yard. After all, the Sears farm catalog had pictures of pigs and horses and sheep all in the same pasture, so why not goats too?

Well, it turned out that pigs aren't the slow-moving, stolid eating machines we'd expected. In fact, at most any size they are active, nosy, curious, and mean. Worst of all, they are nibblers and will chew off the end of anything they can get at, from an apple twig to a goat's udder. The pigs nipped, snorted, and hassled the goats out of the pasture and right between the charged wires before I was halfway back to the barn. So I was forced to neglect the gardens, postpone other important projects, and spend scarce cash and unplanned hours of labor building goat quarters during the wettest, buggiest part of the year. And on top of that was all the time needed to become familiar with the new family members—no books can teach you how much of what to feed and when, or how to make a pig trough that will hold water but not spill. Or that when a goat isn't eating out of its grain trough it's likely standing on it or getting droppings in it. Contrary to popular opinion, goats are about the most fastidious critters there are, and they won't eat from a soiled dish. With all that to learn the hard way, one new kind of stock would have been a job. Two was too much. For a while there, homesteading was very nearly as enervating as the old nine to five job rat race.

Taking It Easy Is Important

This "go slow" admonition applies to just about every aspect of shifting from a city life to a homestead, not just taking on animals. There is a vast difference between the farm boy of two generations ago taking over the family homestead from his Dad, and a modern city type—or even the offspring of a big corporate-type farmer—who tries to get into full partnership with the land from scratch.

Like most of the best modern homestead locations, ours had been a family farm up to the 1930s or 40s. But in the ONE STEP AT A TIME intervening years the fencing, springs, outbuildings, and machinery had been let go to ruin. All we started with was the house and a good sound barn (and a blessing that is), so that everything in the way of farm facilities that the old-time farm families built up over the generations had to be put in from the foundation up. Trying to set up a complete working homestead in a year or two, or even five or six, was out of the question. Not that we didn't want to get it all done at once! But we were both working at money-making jobs and lacked time and energy to take on a great deal at any one time till the financial pressure eased. Now, since we got into the unfortunate hog/goat combination spring, we are taking it easy again.

I suspect that a major cause of failure of so many recent homesteading attempts is that people tried taking on too much at one time. Their enthusiasm overmatched their resources and their expectations exceeded their know-how, time, and ability to get things done. It is just too, too easy to let yourself be misled into thinking that the natural virtue of hewing closer to the land means that some Divine Providence will assure quick and painless success. Not so. Providence doesn't build fence. It's asking for disappointment to plan on doing in a year or two what it took earlier generations whole lifetimes to accomplish. Homestead bulding takes time and experience. Learning is a great deal of the fun of it. So is anticipation—looking forward to broiling those home-grown, organic beef steaks, or swimming and fishing in your own pond, or whatever homesteading pleasure lies in the future.

Running Sheep

Looking back, I'd say the best way to start with large stock is to run a few sheep on summer pasture. (The fact that we didn't start that way doesn't mean that we shouldn't have, it only goes to prove that we've learned a bit.) The years when they come cheap we buy a pair of spring lambs in April or so, run them with the goats till the grass is well up, then keep them in the orchard till the grass goes down. Then they are fattened up on hay and grain and slaughtered in early fall for young mutton and sheepskins. That way we don't have to bother overwintering, dipping, shearing, or worrying about hoof rot, parasites, and all the other problems that make sheepmen in the West so worried they shoot every coyote and bald eagle they can get their sights on for fear of losing a lamb or two.

We get lambs at between three and four months of age, just weaned and prime for slaughter as milk-fed lamb, if such is your intention. Young male animals tend to be cheaper than ewes and they should have been castrated and had their tails removed (docked) before they were two weeks old. The desexing is to keep them from developing into rams with an attendant foul disposition and rancid odor. The docking is to prevent manure from caking in the wool under the tail, which causes all sorts of unpleasantness. If sheep haven't been so treated, I would question the knowledge of the breeder and go sheep shopping elsewhere.

BUYING LAMBS

Lambs, like any living creatures, are susceptible to health problems. You can take one of two basic approaches to keeping livestock fit. The modern way is to dope them up on chemicals and maintain an artificially "clean" environment, preferably huge barns or pens with all-cement floors. The agribusiness interests that promote such management practices are kissin' cousins to the drug makers who fill the magazines and TV tube with ads telling people to spray everything in the house, themselves included, with sanitizers and deodorizers and to take a different pill every half-hour to prevent the sniffles. All as silly as it is unnatural. With all the animals, as with our own health, we take the second approach, figuring that nature pretty well knows best, and it's most natural to give them an environment as close to the wild state as possible. So the sheep, being grazers, are let roam the meadows, pigs are turned out to root in the oak woods, and goats get fed tree leaves and twigs—their natural browse.

KEEPING 'EM HEALTHY

Of course milk goats are native to the Mediterranean area, not New England. Javelinas, the American wild pigs, don't range much farther north than Texas, and the closest American relative of a lamb would be a young Rocky Mountain bighorn sheep. And to keep these creatures in good health outside of their native haunts takes a bit of improving on nature, New England style. This means, for one thing, trace minerals. An absence of selenium in the diet causes "stiff lamb," where a young sheep just stiffens up and dies. A lack of cobalt can cause pregnancy problems in goats and make their milk unpalatable.

Each section of the country has different soil, and each soil will have a different mix of trace minerals. In general, feed stores sell salt licks which contain minerals in local short supply. For example, we need iodine, fluorine, iron, and several others in addition to the two mentioned above. It's not a bad idea, though, to check with the nearest veterinarian who can readily tell you what supplements are needed by which animals and the best way to provide them.

You should learn, by the way, to do your own doctoring for routine animal illnesses. It's a bit of a chore to whip 200 pounds of sick pig into the vet's office, and barn calls are expensive. But don't think you can do without a vet altogether. We started out using the old-time remedies such as tobacco for internal worms and boiled blackberry root for scours—diarrhea in livestock. But after losing a pig, I quickly took up modern medicine. The Merck Company Veterinary Manual, obtainable through most farm supply outlets, gives all the information you need to do what you have any business doing and lets you know when to telephone the vet, who didn't spend the better part of a decade in school for nothing.

Starting the Lamb

THE FIRST DAYS   Like all new livestock, our young lambs are penned alone in the small all-purpose run for several days to see if they have brought any diseases with them. Many young animals will have scours for a day or two just because of the change in locale and feed, but should pass normal droppings shortly, which in sheep are collections of dark spheres like giant rabbit droppings. (If all the upcoming attention to fecal detail makes you a bit squeamish, don't think you're alone. You get used to it, though, and a homesteader has no choice but to become involved in the most basic aspects of an animal's functioning. Anyone who faints at the sight of blood or who would be really bothered by wading up to the ankles in honest animal manure should consider something other than the homesteading life, or at least move on to the next chapter of this book.)

SCOURS   If lamb scours haven't stopped after twenty-four hours,
TREATMENT   and particularly if the animal acts listless, its eyes dull—best not to wait till it stops eating and lies down, by which time it may be too late—I dose it with a broad-spectrum antibiotic and a specially formulated vitamin/high-energy feed supplement. Both concoc-

tions are soluble in water and when mixed with warm milk and a little honey, are a treat readily taken by most young animals. Goats are an exception, especially the older ones. In any event, if an animal refuses medication it may need a drench, a forced drink. You need a drenching syringe; one of those rubber and plastic gadgets used to baste turkeys is fine. The medicine, dissolved in warm water, is sucked into the bulb. Then I hold the animal's head level, *never* raising the muzzle above the horizontal lest fluid

Frankly, sheep aren't too bright. That can be a benefit, for unlike goats, they'll stay pretty much where you put them without elaborate fencing. Pastured, they'll survive the summer without constant attention, which is good for the busy homesteader.

gets breathed in. The end of the tube is poked through the lips in the gap between biting and grinding teeth, and back into the mouth, but not pressing down so the tongue can't swallow. The fluid is squeezed out a bit at a time and you usually have to force the muzzle closed to be sure it is all swallowed. No animal likes a drench, and it is usually a good wrestling match. Two people are better than one and they had both best wear stout, high boots. Even a young lamb has a good kick.

WORMING

Once well perked up, the young stock is wormed. Merck has come out with a new product, Thibenzole or Thiabendazole (TBZ for short), which is a real advance in veterinary medicine. It cleans out most internal parasites with a single dose and is nontoxic enough to be used even on sick animals. TBZ is fine for all herbivores in the barnyard. Hogs are different, needing another wormer. We use Atgard as of this writing.

Two days after worming the lambs are turned into the goat run, despite the temporary complaints of the permanent residents, and I clean out the small pen. If there were any worms present it will be obvious: they look like week-old spaghetti. All the accumulated droppings are burned and feeding equipment is scrubbed well with disinfectant. I let the sun do a month or so of disinfecting on the soil, since it's better at that work than anything else, and just before the next shift of animals goes into the pen, I toss in a new layer of fresh bedding hay.

Pasturing the Sheep

Once they have a little age on them, sheep are the most placid and easily managed of animals. To put it another way, sheep are plain stupid. So long as ample grazing and water are available, they won't climb even a low stone fence and a single strand of electric fence will hold them if they've been trained to it. (Details of the training process and putting up electric fence are given in the section on hogs.)

TWO DANGERS

Before pasturing out sheep, there are two serious dangers to consider: dogs and poison plants. It is a rare lone dog that will do much more than run sheep, barking and perhaps nipping at their hocks. But two or more, even your own house pets, can revert to ancient pack instinct and will kill sheep and goats. We never let more than one of our dogs run loose at any one time. The German Shepherds sometimes yap around after the sheep, but have never hurt them. And our dogs tend to keep other hounds away from the place altogether. For the occasional pack of strays that comes night-visiting, I have a strong floodlight trained on the pasture and a twelve-gauge shotgun loaded with rock salt.

Much more common, dangerous, and harder to control

Mountain laurel

than dogs are the poison plants. The worst are members of the American laurel group, mountain laurel, etc. They are nicknamed lambkill, pigkill and similar names, for the obvious reason. They are easily identified, being just about the only evergreen shrub around with broad, flat and shiny leaves, not needles. The plants are a yard or so high, with a sparse leaf cover. They usually grow in groups, always in partial shade. The laurels simply must be grubbed out. Just one or two leaves can kill a lamb, pig, or goat.

Some ferns are a mild danger, but usually prefer dense

Some homesteaders raise sheep for their wool, spinning it into yarn for knitting or weaving. Louise and I don't have so much spare time that making homespun is a worthy endeavor.

shade where you wouldn't pasture sheep anyway. Loco weed is a problem in the West, and water hemlock occurs in marshy thickets in the northern half of the United States and into Canada. This is the same poison hemlock that killed Socrates, not the familiar hemlock tree, properly called yew. The poison plant is a member of the parsley family with a root that resembles a sweet potato, smells delightfully of parsnips, and has killed many a pig and probably a few readers of wild food manuals, but no sheep since sheep don't go rooting around foolishly in swamps.

The most common poison plant is the wild cherry tree. Young shoots and withered leaves can generate prussic (hydrocyanic) acid in a ruminant's stomach and if eaten in sufficient quantity, it can kill. The animal may bloat (swell in the gut), bleat from pain, froth at the mouth, or leap around aimlessly. It should vomit, which is desirable, and a good drench of epsom salts or a mouthful of dry bicarbonate of soda can help the process along. Even better is to check pasture each spring and dig out any young wild cherries. The twigs are a shiny dark mahogany color, the bark smooth with little pits in it, and when broken a branch will smell strongly of wild cherry cough drops and have a strong and bitter wild-cherry taste.

Wild cherry

Water hemlock

If you dig the young plants it is essential to get all the root structure or more lethal little sprouts will come up. Unless you want to spend every spring for years digging sprouts. don't cut down any good-sized wild cherries. They won't send up sprouts unless cut, and if you get the sheep out of the pasture before leaves come down in the fall, they won't cause any problems.

Sheep are nature's lawn mowers, and though we don't appreciate the droppings around the house, they have more or less free range elsewhere around the homeplace, particularly in the orchard. They keep the plants under the fruit and nut trees well cropped, eat up the deadfalls, and their droppings and urine do a better job of fertilizing orchard land than any amount of chemicals could.

FEEDING

Sheep apreciate a bit of grain each day, and fresh, clean water should be available at all times. At the end of their range closest to the house, the lower end of the orchard where the gardens start, I keep a shallow wood trough. At morning chore time, unless it is raining, I take up a small bucket of goat grain or cracked corn. Their water supply is a fifty-gallon oil drum set on a rock pile in the shade of an apple tree. The bung is replaced with a stock-watering cup, a small basin with a paddle-valve the animals can nudge with their noses to get a drink. The drum is filled every week with water from the house pump.

With their feed and water near the house, the sheep make that area their preferred bedding spot and will spend nights in the pine and maple groves at the orchard corners. So far I've never seen any problems from their being exposed to the spring and summer weather. The thick fleeces that make such fine, waterproof coats for us must do the same for the sheep while's their turn to wear them.

HOMESPUN

I'm often asked why we don't overwinter sheep, increase the flock, and shear, process, spin, and weave the wool into cloth. Making your own homespun has a fine old-timey ring to it and I guess that's the ultimate in self-sufficiency. But it is unbelievably time consuming. A neighbor who does the whole sheep and wool thing estimates that her time would be worth only a few pennies an hour if she sold her output, which she wouldn't think of. Her family and friends wear it. But then she loves her sheep and the hum of her spinning wheel, the clank of the loom, and the feel of lanolin-rich wool moving through her fingers. It's a genuine pleasure we can't deny her, but one we don't share.

Getting that far back into pretechnology is fine, but I'd

have to call it recreation, though productive recreation in the same league with my fishing or Louise's pottery. The cash value of time spent is strictly secondary to the pleasure obtained, though the money *not* spent (on wool sweaters, ovenware, or fresh trout) may in time pay for the equipment needed. Still, any of the three homestead-style hobbies are productive and vastly more satisfying than playing golf, staring at pro football on the television, tooling around on snowmobiles, or many other more "civilized" pastimes.

Who Fed Henry Thoreau?

From time to time we are reluctant hosts to groups of (usually very young, city-bred) purists who sometimes question the integrity of our commitment to a "back to nature and the land" existence. After all, they say, Henry David Thoreau didn't have electricity and central heating at Walden Pond, and we *should* make our own cloth, use horses to plow the fields, and etc., like the original pioneers. To give these folks a factual perspective on the life actually lived by the early settlers, I take them to the little cemetery down the road. There stands the headstone of a farmer of the late eighteenth century who had no choice but to provide nearly all his needs himself. Integrity he had, I don't doubt, and he lived well into his eighties. But he is surrounded by the graves of several successive wives and children of all ages who just couldn't hold up to the savage poverty and privation, the endless hours of back-breaking labor of a life totally powered by the muscles of men, women, and their livestock.

Face it, there was at least as much bad as good in those good old days romantics always moon about. But the good of those times is worth reviving and combining with the good of life in late twentieth-century America. It's a continual search and compromise, but well worth making in our opinion. For example, when I have a common cold, I find that a steaming mug of old-fashioned rose hip tea with honey is all the medication I need. It's rich in natural vitamin C, the steam soothes sore nasal passages and sinuses, while the astringency and sweet helps a scratchy throat and cough and settles a jumpy stomach. It isn't any cure—only time is—but it beats Bufferin, Anacin, speedy Alka Seltzer and Bayer aspirin any day. On the other hand, if the baby's temperature ever tops 101°, I'll be thankful for all the technology that enables me to drive to a hospital where an MD with half a lifetime of training under his belt has a whole arsenal of modern medicines at his command.

Oh yes, about culture hero Thoreau. His two-year stint at Walden was book research. A lovely book, I'll admit. But when the good ladies of Concord town forgot to bring a really big lunch hamper out for their afternoons of pondside Transcendental chat, Henry hot-footed it into town to beg a meal off Ralph Waldo Emerson. Now, that's history pure and straight, with the romantic nonsense edited out.

Hogs

Speaking of romantic fiction, I suspect that Walt Disney, Inc. or someone else in the animal nonsense business has been spreading false rumors about hogs. More and more city visitors of late are telling us how surprised they were to learn that hogs aren't really the horrible creatures of fable, but actually clean, sanitary, and lovable animals. They aren't as surprised as I am. Now, our son Sam, having been raised with the real porcine article slobbering around in the back—the *back, back*—lot, has no illusions. One of the first terms he learned to say clearly was "Horrible Pig," which is the name assigned to each Penelope or Olive Oink as she grows from pighood to hogsize. Indeed, when Sam has a cold, Louise has no problem getting him to blow. She only has to ask him, "How does Horrible Pig go?"

HORRIBLE PIG

I'm going on in this vein because I don't want to be accused of luring someone into hog raising on false principles. Hogs are simply ghastly, no two ways about it. I mentioned earlier how they scared the goats. Well, they scare people too. The electric fence is hidden by brush where the pig yard borders the road, and more than one innocent hiker has been sent scurrying by what appeared to be 200 pounds of unrestrained wild boar bearing down on him. Hogs don't go "oink, oink." They snort and growl—one passer-by thought she was being attacked by an odd-colored bear. The hogs are trained to the electric fence and would never actually get out to the road, but not everyone knows that.

On a more sober note, but one that isn't widely known, a grown hog will kill and eat just about anything it can get. This includes very small children. So our back yard fence is extra strong to keep the tinies out of the woods. And children through grade school age are strictly forbidden to approach the hog lot without an adult along. There are few ironclad rules on our place, but that is one of them.

Another piece of nonsense I've heard from folks who learn about hogs from TV programs is that the animals are

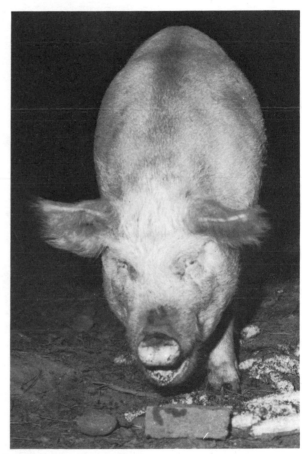

Horrible Pig is so called because the name fits. We feed Horrible Pig a lot of stuff that ordinarily would go into the compost heap or out for the garbageman. But the garbageman doesn't stop here, and Horrible Pig processes garbage as well as any compost heap.

naturally sanitary, selecting a corner of their pen or yard for a latrine. This is more or less true of small pigs kept in a small pen. But as they grow in size, hogs lose their manners. An adult hog produces up to its own weight in manure each month, some in a latrine spot, but most where the urge happens to hit. At the feed trough (or in it), in the mud wallow, etc.

Since hog manure comes in fairly dry, firm hunks it is easily removed so that a daily shoveling will let you keep the animals in a small space. Without such stringent sanitation, preferably on a cement floor that must be hosed frequently, small quarters are usually out of the question. You could conceivably keep a hog in a small pen by pitching in a daily layer of leaves, much as we deal with the chicken run, and at the end of the year you would have perhaps a ton of good, largely composted fertilizer. We find it easier, though, to give the hogs a more natural environment, letting them run in about a half-acre of woods. A stream runs through the trees so they have plenty of pure water. There are trees to rub against, forest loam to root around in, and even a little pond that dries out to a two-hog-sized mud hole when the weather gets hot. So the hogs are happy, if horrible.

Starting With Hogs

SWINE AND BREEDERS

I'd say the first rule in getting pigs is to obtain only top-quality stock from a reputable breeder. The second rule is not to take children with you when picking the animals up. We learned both rules the hard way. I got our first pair of pigs by answering a newspaper ad and ended up driving such a long distance I felt we had to come home with animals to make the drive worth it. And I took along a batch of tender-hearted kids. Their faces fell even farther than mine when we found the breeder was a garbage collector who kept little pigs in a wrecked car at one corner of a place that was nothing but one foul-smelling pig sty.

Now, I've nothing against garbage-feeding hogs. I do it myself; and in fact, a good bit of the pork that appears in stores has been raised in large part on store and restaurant wastes—recycling in the best sense of the term. However, the garbage should be cooked, the leavings cleaned up, and hogs kept in sanitary conditions for a commercial operation to appear and smell anything but awful to a casual onlooker, as well as to satisfy laws in most states. From the look of him our first hog-seller didn't much care for any laws but his own and I was glad to get us all well upwind of his place as fast as possible. Needless to say, subsequent arrangements for pig replacement and breeding have been done after a careful check.

Those first pigs went into the all-purpose pen for the

routine disease check. They both had bad scours, and as mentioned earlier, only one survived very long. The other seemed contented enough, ate well, but kept drizzling despite treatment with several old-style cures. Finally it swelled up like a football with legs and died from what turned out to be internal worms that had gotten too good a hold when the pig was tiny. That was when I switched to modern veterinary medicine.

Our hogs aren't of any particular breed. Indeed, I think **BREEDS** that crossbreeding tends to produce healthier stock, as is true with hybrid plants. And these days, breed really doesn't make much difference. Time was, hogs were bred for their lard but the fat lard-type hog has just about disappeared, to be replaced by the modern long-boded, well-fleshed meat-type Yorkshires, Hampshires, Poland Chinas, and other breeds peculiar to different parts of the country. The coloring and confirmation of purebreds are important to 4-H-ers and commercial hog farmers, but for the homesteader's porkchops and hams, any good hog will do. I'd say, just get the best pigs you can find. Try to get the pick of the litter, choosing the biggest pigs of the lot; they should be about thirty-five pounds at weaning age of about sixty days.

If planning to raise your own pigs, get females and be sure **SEX** the vulva—the external sex organ which is the little mubbin under the anus—is well developed. If you get males, be sure they've been castrated. It isn't worth feeding a boar unless you plan to go into the pig breeding business. Besides, boars get very large, have sharp tusks and mean dispositions. I wouldn't have one without a chain link fence to hold it. Come to think of it, I wouldn't have one even with such a fence.

Hog Fence

"Hog-tight" is the old term describing the strong fencing needed to physically restrain adult hogs. With a strong rooter at the end of a powerful neck and shoulders, a hog can quickly work its way under any weak fence. Then too, they delight in scratching on anything handy, and two to three hundred pounds of hog leaning on a fence and scrubbing back and forth will tax any but the strongest posts and fence wire.

Vastly cheaper, easier to put up and maintain than steel **FENCING** fence is an electric fence, a few strands of electrically-charged wire with a sting that is harmless but effective if the animal is taught to respect it. You can buy fence-chargers in either battery-operated or 110-volt line current models for twenty-five dollars and up. The fence is nothing but galvanized steel wire; for hogs I use two strands, one a bit less than a foot off the ground and the other a foot above that. The wire is fastened to posts and trees with ceramic insulators shaped like big thread spools. These are nailed in well with special twin-headed spikes, and the wire is fastened to the insulator with a special wire clip. You can get two-piece insulators and crimp the wire between the halves. This makes for easier wire stringing, but water will get in the crack each good rain and your fence will be shorted out more often than it's working.

The electric wire needs only to be pulled taut by hand. However, to provide an added measure of security I also put in a strand of barbed wire between the charged lines. This was stretched with a rachet-equipped block and tackle rig—mainly a fence-stretcher but also used to haul up large livestock at slaughtering time. The barbed wire is simply stapled to the trees, and should last a good twenty years. The electric fence will probably rust out in ten.

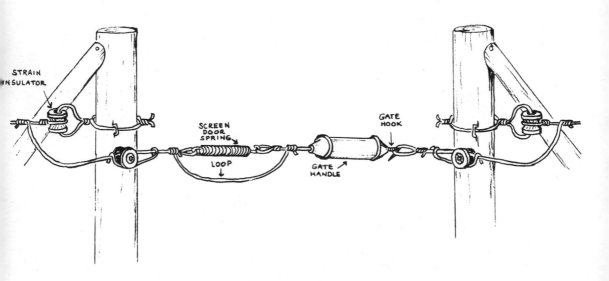

The fence is low enough that an adult can step over it, but I also put in several openings at strategic points. You cut the wire, put a loop on one end, and on the other fasten a ''gate''—a plastic tube with spring-loaded steel rods running through to conduct the current. Since the plastic doesn't conduct current you can pull the gate back on itself enough to compress the spring so you can pull the hook end free and walk through.

Any large farm animal could walk over, under, or through an electric fence if it wanted to, so they must be trained to fear the wires. Once trained they will stay inside the fence even if the current is turned off most of the time (though I leave ours on all the time mainly to keep dogs out; it uses very little electricity).

FENCE-TRAINING LIVESTOCK

To train an animal I arrange a length of wire in the all-purpose pen, at about nose level for the beast being trained. Then some tempting food is placed on a platform just on the far side of the wire and the poor creature zaps its moist nose over and over again till it learns the lesson. With young pigs, a couple of training sessions in the all-purpose pen are followed with one or two more out in the hog yard after they are transferred, and that's that. They will continually test the wire till they reach full size, so don't turn off your electric fence while running small pigs.

Hogs Housing

Hog housing isn't really much of a problem; the lard of big ones protects them from almost any degree of cold, though winter-born litters need heat. Our adult hogs don't even seek shelter till it's around ten degrees. The main problem is heat. Hogs must be protected from direct sun, especially in hot, humid seasons. Ours can run in under the trees or cool off in the stream or hog-wallow. Running hogs on cleared pasture, you must provide shade; a couple of hay bales with planks over the top will do, though eventually such a house will be snooted over. I build our hogs a hay and plywood winter shelter up against an outbuilding. The bales are pinned to the earth with stout poles and covered with a plywood and roofing-paper roof. It usually stands up all winter.

HOUSING

While in the all-purpose pen, the first lot of pigs was fed and watered in troughs. Even little pigs are rooters, snooters, and walkers in their feed and it took me some time to come up with a

FEED TROUGHS

Our rudimentary but servicable winter hog shelter

Feed trough

design that was relatively spill- and damage-proof. I ended up putting three yard-long boards about eight inches wide in an "N" shape with stout two-foot-long boards nailed at the ends and the whole caulked with a nontoxic putty. In effect it is two troughs, one on each side, and will hold feed or water in whichever side the pigs have rooted it onto.

PIG FEED   The old-fashioned way of feeding small pigs was to mix up cornmeal and milk into a slop that rapidly turned sour if not cleaned up immediately. Today, farmers keep clean water available and let the pigs self-feed a dry ration. "Eating like a pig" is a saying that implies gluttony, and that's not fair to pigs. Their trough manners are awful, to be sure. But no pig will eat more than it truly needs. However, going from zero to two hundred pounds in six months takes a lot of food. We hosed fresh water into the pigs' trough several times a day and fed them three times a day—household garbage plus corn or a mixed grain ration, and as much of it as they would clean up in a feeding. They also got about a pint of goat's milk apiece in a bucket twice a day.

A pig's digestive apparatus is similar to yours and mine, which is to say that it thrives on a mixed diet. In the wild, pigs live largely on nuts, berries, roots, and acorns—"mast" is the correct term. They also enjoy any grubs or other animal food alive or dead they root up, and if our hogs' behavior is any indication, they also nibble at small saplings, toadstools, and just about anything else that's palatable and handy.

Commercial raisers have hog feeding down to a science, giving small pigs a grain combination that provides 16 percent protein, then reducing it to a 14 percent ration for grown hogs. With good, fast-gaining stock and a computer-designed, chemical-infused feeding program, some growers have gotten a 3:1 feed efficiency ratio: having their animals put on one pound of weight for every three pounds of feed eaten. The old-time ratio was 7:1: seven pounds of feed to produce a pound of pork. I suppose our hogs are seven-to-oners. Of course, the majority of their feed is household, garden, or other "waste." On average, I'd guess that every pound of ham or headcheese was fed on four or five pounds of free "waste" and one or two pounds of grain feed, which we bought in the early years and now grow practically for free, the way we do it.

I base this estimate on offhand observation, not any careful records. But each hog should have a good six pounds of feed a day. I weigh the day's accumulation of scraps, and if they don't come up to six pounds per hog, I add in corn or poultry mix. On average, I'd guess that the grain makes up a third to a quarter of most feedings.

At first we tried feeding grown hogs in troughs, but they got pushed all over the lot and eventually torn apart. Now we have a half dozen spots along the fence where I just toss the feed in. The animals root around after the food in the leaves and pine needles, in the process ingesting enough good soil and organic matter to provide the iron they need plus all the trace elements but iodine, which is supplied in the salt we use in cooking our own food, and that they get in the garbage.

KEEPING 'EM HEALTHY

With such a varied diet, plenty of pure spring water and lots of exercise space, the hogs are active, healthy creatures, if horrible. About the only routine medication they receive is worming powder, sprinkled in a feeding anytime an animal is moved from one pen or pasture to another. Nature just naturally takes care of sanitation, though I pitch a good layer of leaves on the more obvious latrine areas in a particularly wet spring when a wind from the direction of the pig lot brings with it a reminder.

Occasionally an animal will "go off its feed," which means it stops eating and lies down. Little pigs so afflicted need quick attention and the same sort of vitamin and antibiotic combination used on small sheep works for them. So far, our grown hogs have cured themselves. They'll lie down and refuse feed from time to time, but will get up and wander off unhappily if nudged with a boot. The problem is most likely indigestion and clears up in three or four days. I strongly suspect, for example, that our last such spell was caused by feathers; I'd given a hog an old rooster that died of the miseries when a young male bird beat him out of his top spot in the flock. The hog ate it feathers and all and was down the next day, though up again after twenty-four hours.

I suppose that some day we'll have a grown hog die on us. If I can't find a rendering company to come out our way and haul off the carcass, I'll have to resort to another old-time solution: get out the spade and dig a hole on the down-hill side of the late lamented, hoping not to encounter any too-big rocks, then roll her in and cover her up.

Hog Breeding

If not buying feed is the first step in growing economical pork, not buying baby pigs is the second. This means breeding a sow once or twice before slaughtering her and it's another dimensional step forward in animal raising. You have to locate a boar, get the sow to him and back, then deliver, raise and sell, or fatten up five to a dozen pigs. A sow's first litter is normally her best effort, and we usually slaughter sows after they have weaned the first, typically having eight or nine good pigs surviving if all goes well. One warning: many novices see each litter of pigs as an easy two or three hundred dollars. If a market for suckling pigs or weanling feeder pigs exists in your area, you may be able to sell most of each litter, but don't think those hundred dollar bills will arrive automatically.

SOW IN HEAT   First problem is knowing when a sow is in heat and amenable to a boar's attentions. Starting at about six months of age a young female (gilt) comes into season every three weeks for a two-to-four-day heat. The only visible evidence, except to another pig, is a swelling and reddening of the vulva, accompanied by a whitish discharge from the genital opening. Some sows will show season by scratching their rumps more than usual or will carry their tails higher than normal. Others will try to mount other females or act more aggressive than usual.

It's best to record the start and duration of several heats to be sure of the gilt's cycle, then get her to the boar several days before a heat is due to commence. If the boar's owner is a good hog man he will take care of the whole process for a fee that should be around ten dollars. There are all sorts of problems in getting hogs of different sizes and weights mated successfully—you need breeding crates and such—that I don't know anything about. The fellow who breeds our hogs has been doing it for twenty years and I'll leave it to him. He breeds a gilt twice on successive days, claiming that it increases the litter size by one or two pigs.

SERVICING

Hogs will accept service pretty much year-round, so the .timing of the whole process is up to the homesteader. For selling pigs, it's best to breed around the first of the year. After a four-month gestation period and two months of nursing you will have pigs available in June when most people want to take them on. We prefer to time breeding for slaughter, which is best in early winter when bugs are gone and a few inches of snow on the ground simplifies everything, from cooling carcasses to keeping the meat clean. So we try to breed in mid-May, farrow (that's pig-talk for birthing) in mid-September, then slaughter the sow and excess suckling pigs around Christmas.

Pigs aren't your daintiest critters. I built this roofed feeder for the young pigs, and they use it for shelter. When they get older and are moved to the electric-fenced pen, they'll get their food on the ground.

Breeding Mechanics

The worst bother in the whole process is getting the gilt to the boar. Hogs won't be driven up ramps or into trucks like cattle, nor will they follow you like a sheep or goat. Slaughterhouses move hogs through narrow and very strong chutes, using electric prods to keep them moving. Two or three people with strong poles or flails can back a hog into a crate but we've found the best mover to be an overripe banana. I back our old farm trailer up to the pig yard, attach a stout ramp to the back, and coax the animal along. She would be reluctant to move out of her accustomed area so I push the cart into the yard. Then it takes a few minutes of banana and stick to coax her into the trailer. Getting her to leave the trailer and return to it at the breeder's place usually takes a bit more force, and more than once it's taken one man with a lasso pulling and another with an electric prod at the other end.

PRENATAL FEEDING

Once back home the pregnant hog requires added nutrients, the total ration gradually increasing to eight pounds over the first three months of the four-month term. As the garden begins to produce, hogs get increased amounts of radish tops, asparagus bottoms, trimmings from spring greens, along with the kitchen leavings and corn. As the weather warms so does the root cellar, and as potatoes begin to sprout, apples to soften, and carrots to begin putting out little yellow fronds, this excess of the prior year's harvest goes toward the current year's pig crop.

I'm particularly careful to include a good protein source in each meal during pregnancy, but I try to keep the carbohydrate level low. The sow should be kept as trim as possible, indeed should be permitted to lose a fair amount of fat, to make carrying and farrowing the litter easier. I've heard that Canadian farmers keep their hogs especially lean by feeding them oats rather than the U.S. corn diet. We may try that some day.

AND HOUSING

On a day in the beginning of the sow's sixteenth week I worm her, then the next morning coax her into a separate area about twenty feet on a side against one wall of the pig lot outbuilding, also fenced with electric wire. By mid-September the weather is usually cool and dry enough that she will be mud-free and clean by hog standards. (Both the U.S. Department of Agriculture booklets and several recent books on homesteading blandly tell you to wash up a farrowing hog. But they don't bother telling you how.) However, with the large yard and woodsy habitat our hogs enjoy, any caked mud would be just that: honest soil. And that naturally wears off during the several prefarrowing days the sow spends in the small pen. While there she is fed and watered in troughs and manure is cleaned up pretty much as it is produced, generally during or right after feeding. This extra contact between the hog and homesteader also serves to prepare her for human presence during the farrowing itself. Can't say we get to be good friends during this time, but we do learn to tolerate one another close up.

I pitch a bale of old hay into the pen along with an assortment of small saplings and branches. As her time approaches the pig will make a nest in one corner of the pen. The last two days or so she gets a reduced feed ration and if any signs of constipation appear, a warm bran mash as a laxative. Then, usually towards an evening as a chill rain begins to fall, she will begin milling about restlessly, then lie down in the nest at the least convenient angle possible, and begin producing little pigs.

Pigwifery

Though both the weather and the hog's disposition may be vile, the birthing must be fully attended if the homesteader expects to salvage more than one or two pigs from a litter. Not that pigs are bad mothers. On the contrary, they can be savage in protecting their brood. However, two-hundred-plus pounds of hog can do a lot of damage to a litter of tiny pigs in the process of simply rolling over. So I attend the farrowing.

As each piglet pops out—with such comparatively small young, hogs seldom have birthing problems—I take it, remove any membrane or mucous, particularly around the mouth and nose, snip the umbilical cord to about a two-inch length, and dip the end in purple "horse medicine." I then dry the piglet with a soft towel, give it a squeeze or two if it isn't breathing already, then nip the points off the eight little tusks with side-cutting pliers. This is to keep it from wrecking the mother's teats.

FARROWING

Next I rinse any dirt from the sow's belly and hold the pig up to a teat long enough for it to get a first meal. The first milk is a thick fluid called colostrum. It's nature's protection against disease, providing the young with antibodies which protect them till they are old enough to produce their own. Without the colostrum, the litter would almost certainly die within a few days.

Once fed, the pigs go into a heavy wood box some five by two feet in size and two feet high, open only along one long side. This is filled with plenty of hay and will serve as the piglet's bed and refuge from the mother till they are large enough to fend for themselves.

POSTPARTUM PROCEDURES

Once a good half-hour has passed without a new pig appearing, I leave the new mother—now an official sow—to clean up the afterbirth. This reduces her appetite for a day or two but is nature's way of restoring to her as much of the nutrient-rich birthing apparatus as the pigs don't need. Once sow and piglets have rested up from the births during the remainder of the night, the litter will come squirming out to feed. Early the next morning I try to get there before they come out and arrange the nest and pig box in a corner of the pen most protected from the weather, the

A half-grown porker, trained to keep inside the electric fence, getting a treat of overgrown zuccini.

box up against the outbuilding wall so the sow can't move it easily. Then for the next few days I make frequent trips to the pen, and give the sow small feedings to let her adjust without constipation. It also distracts her so I can keep putting the piglets back in their box. This, plus keeping the nest area bedding fresh, deep, and well fluffed, keeps too many pigs from being squashed.

Occasionally a sow will not come into milk and quite often one will have more pigs than she has functioning udders. The traditional farm practice is to spread the extra young out among other brood sows. However, with only one litter at a time, we have to bottle-feed. If we don't, the sow will eat the pigs. So it's baby bottles with fresh goat's milk morning noon and night for a week. After that time the pigs can navigate well enough to be fed from a pan, though you have to nearly drown them to get them drinking and more milk is spilled than drunk. All in all it's probably easier to lose a few pigs from maternal neglect or misconduct than hand-feed.

In two weeks' time the pigs are getting around well and can be put on feed. I remove the electric fence between the farrowing pen and hog lot and coax the sow out with a good feeding thrown out at the far side of the lot. Then I replace the fence with a foot-high board the sow can get over to nurse, but which keeps the pigs penned for several weeks more. Then, twice a day while the sow is distracted by a half-feeding some distance from the pen, I feed the litter. First feed is stale bread soaked in milk. In a day or two we add in the softer kitchen scraps, then a bit later, ground corn soaked overnight in vegetable cooking water. This mash is slowly phased out so that at two months of age the pigs are eating the same ration as their mother.

It's during feeding time, when the sow is distracted with her own meal, that I castrate the males. This can be done at any time after the testes have descended into the scrotum, but the earlier the better. There are a number of ways to get this job done; I use a gadget called an Elastrator. Obtainable at any farm supply outlet, it stretches open a tough elastic band. You just slip the scrotum through the elastic and let it draw tight by closing the tool's jaws. The band cuts off circulation and the scrotum will dry up and slough off without a squeal of pain or drop of blood.

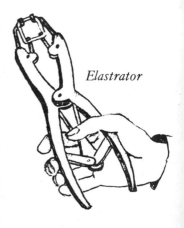

Elastrator

Little pigs are subject to nervous disorders, believe it or not, so it's best to space the traumatic things you do to them at least a week apart. I usually give shots (as the vet suggests, hog cholera being a possible danger in most areas of the nation) at a month of age, worm at five weeks, castrate at six, and begin selling and thus weaning pigs at two months. Commercial breeders would take the whole litter at sixty days and rebreed the sow immediately. We just leave the pigs to run free in the pen after two months in the farrowing pen till they are sold or slaughtered as suckling pig.

A final note: when handling little pigs, especially when removing them from the pen, wear stout boots, heavy gloves, and keep an eye peeled for the sow. One piglet squeal will bring her running in a rage of maternal protectiveness that just might ignore an electric fence. To reduce the danger of attack, I cover a pig's snout with one hand and step quickly. Once put down it will quit trying to squeal and the sow will never know it is gone.

Queen of the Place: A Dairy Goat

9

High point of the early fall on our place is the arrival of the newest aristocrats of the homestead, the dairy goat kids. Note if you will that I'm not lumping the goats along with pigs and sheep in the chapter you just waded through. That's because goats are different from other farm animals and getting one is, in our opinion, the ultimate step in homesteading. It's a step that should be more carefully thought out than any other, including taking on a cow or steer or a horse.

The reason is that goats invariably wangle a way to become real family members, pets in spite of the homesteaders' best efforts to view them as dumb animals. First of all, they aren't dumb in either an intelligence or communications sense. They have high animal intelligence and a range of bleats, butts, and snorts that convey their requirements most effectively. They are beautiful creatures with great, limpid eyes and dainty limbs, and each has a distinct personality and a demanding but engaging manner that is truly almost human. And on top of it, to have milk you must breed your does, which means that each year you will have from one to four of the most appealing animal babies you'll ever see. It is awfully easy for a careless or warmhearted homesteader to end up an unintentional goatherd.

The trouble with goats; they get to be pets.

So, before buying that first goat, determine that you have a market to sell the young as potential milkers and can steel yourself up to butcher the young bucks for meat and hides. You should also be prepared to butcher a nonproducing older doe after years of faithful service—and be able to enjoy the roasts and chops into the bargain.

Louise made it very plain that her participation in the foregoing aspect of the homesteading experience would be restricted to cooking, serving—and trying to enjoy—the meat. So, after making the hard decision that I would and could do the rest of it, we set out on an even harder task, finding a goat. Real goat people, the breeders and large dairies and milk users, are a close-knit group and hard to ferret out. There always seem to be one or two small-scale goat keepers in each rural community, but we weren't able to find one with animals that were good quality, young, and for sale. Finally I subscribed to the Dairy Goat Journal, a little monthly magazine that only costs four dollars a year (address P.O. Box 1908, Scottsdale, Arizona 85252). It contains fine articles on goat husbandry and health, good recipes for milk, cheese, and meat, plus a lot of material for breeders and show people. The breeders' ads tend toward posterior photos of milking does with such comments as "Udderly Amazing," but there is also a lot of equipment for goat keeping *and* the identity and secret address of the nearest goat breeder.

STARTING WITH GOATS

It's reputed that goat people are a lot like horse traders and I suppose there are some who will gyp a beginner. In our experience though, goat fanciers are good people, though a bit peculiar. Maybe "different" or "individualistic" is a better description of them than "peculiar", and maybe I'm letting on more than I should about myself. But odd as we might be, goat people enjoy talking goats, even with a raw novice. It does help if you know a few goat terms though, so here's the basic information I wish I'd had when I first went off to try talking a wonderful woman goat keeper into selling me one of her family.

BASIC INFO

Female goats are called *does*, the males are *bucks*, the young *kids*. Only the best purebreed bucks are kept for breeding: the others are castrated to become wethers. They are kept for pets, perhaps to pull a child's goat cart, or are slaughtered for hides and meat, which is called *chevon*. There is a lot of information that everyone knows about goats that is dead wrong, for example that goats smell bad. Mature bucks do emit a strong musk from glands behind their horns, especially in the fall peak breeding season. The glands may be removed surgically if the bucks' owner objects to the odor. Most people don't like it. I like it and so do most goat people. But does, kids, and wethers have no musk and if kept groomed and in reasonably clean quarters they smell less than any farm animal I know—and that includes chickens and rabbits as well as large stock.

Everyone also knows that goat's milk tastes peculiar. FACTS AND Nonsense; it is indistinguishable from cow's milk if handled FALLACIES properly in the milking and if the doe is healthy and fed properly. More common "knowledge" is that goats eat everything, including cans. In actual fact, goats are the most fastidious eaters in the barnyard and will starve rather than touch grain or grass that is spoiled, just slightly tainted with excrement, or nuzzled too much by other goats. (I have to empty "refused" hay from the mangers at least once a week.) They will nibble, worry, taste, and play with just about anything from a tin can to your shirtsleeve, or the bandanna in a back pocket. But when it comes to food they are finicky.

Goats are agile animals that constantly challenge our ability to fence them in. We usually allow them to roam in the yard, which is surrounded by six-foot high fencing installed specifically because of them.

Price and Quality

Goats sell from fifteen dollars for an old female whose milking days are about gone, up to many hundreds for a pureblood, high-milking show champion. No matter what the bloodlines, don't expect to pay less than a hundred dollars for a good milker in her prime. A doe kid or yearling that hasn't freshened (begun producing milk) will go for fifty to seventy-five dollars and up, a wether for twenty-five to fifty.

Most goats will grow horns if you let them, but you shouldn't. Goats butt one another in fun and in arguments, and good breeders stop horn growth when kids are tiny. I'd advise strongly against buying horned goats and would advise even more strongly against running horned goats with hornless, defenseless animals.

BREEDS   Goats come in several breeds, each developed in a distinct part of the world centuries ago. The Asian breed, Nubians, are large by goat standards, have a rounded Roman nose, floppy ears, and a black coat with white spots. The Swiss breeds that Heidi herded include the all-white Saanans, multicolored French Alpines, and the most popular breed in the United States, the

French Alpine

Toggenberg

Toggenbergs. A "Tog" looks like a deer, with a lovely brown coat, regular face, and rump markings. There are several other popular breeds including the domestic development, American LaMancha, and Angoras which are raised mainly for their silky hair.

If a goat is a pureblood you can get a pedigree, register it with the American Dairy Goat Association, and enter it in goat shows. However, none of this will guarantee you a good milk supply. Indeed, many U.S. goat breeders select for type—trying to get animals which conform to esoteric points of physical shape—rather than what's really important, quality, quantity, and length of kid and milk production. All of which is silly in my book. **OUR CHOICE** So our goats are what's called "Grades," mixed breeds, though the sire is normally a purebreed. A goat without any pureblood parentage is called a common goat; many of these are uncommonly good milkers.

The first goats we bought had the Toggenberg coloring and a good Nubian flop to the ears. I guess we tend to stick with what pleased us at first, since I alternate between Nubian and Tog bucks. Perhaps someday I'll keep a particularly pretty buck kid from a good-milking doe and start a new breed, Vivian's Toggians. Has a good ring to it, don't you think? Maybe if it turns out, I'll try to sell you one someday.

Saanan

Nubian

Goat Quarters

**PROPER
QUARTERS**

Wild goats are fairly hardy creatures and can get along anywhere from tropical islands to high mountains. Modern dairy goats, however, are the result of years of selective breeding and while still quite healthy, they need proper quarters. Basic requirements include a good-sized exercise yard, preferably offering both sun and shade and having a strong fence. Inside stalls should have a thick layer of clean bedding, solid roof and walls, and doors and windows that close tight enough to keep out any drafts.

Ideally the goats should be able to move freely from sun to

With livestock, nothing goes to waste. Here the goats are getting sweet corn stalks. The kernels go to the family, the husks to the rabbits, and the buttery, gnawed-over cobs to the pigs.

shade, outside to inside shelter, at all times during the day. Nights and on cold days they can be shut inside ample-sized stalls. This too runs counter to something else that everyone knows—that goats can be kept in tiny stalls just big enough to stand in, either tied up or with their necks in stanchions, wooden frames so narrow the goat can't pull free. This bit of foolishness comes from a misreading (of a not-very-well stated section) of a widely available but unreliable handbook on goat raising. And sure enough, it's faithfully repeated in most of the recent homesteading books, written by city-based people who get their (mis)information mostly from the library. Now, it's fine to tie or stanchion a goat to feed, groom, milk, or otherwise service her. But anyone who tried keeping goats so confined constantly would have a long wait for their milk. Lack of exercise—plenty of exercise—during a doe's final month of pregnancy brings on a combination of pregnancy diseases, any one of which can be fatal. Actually, anyone who planned to keep goats stanchioned or in a tie-stall would have a hard time getting animals in the first place. I wouldn't sell him one and I don't know another goat keeper who would.

STANCHIONS

On our place the goats' winter quarters is a horse stall something under one hundred square feet in floor area. This gives the two or three animals we overwinter plenty of space to have their own sleeping and cud-chewing corners plus enough room to mill about in. On warm winter days and then full-time once the biting bugs decrease in spring, the goats stay in the lower barn yard. The yard, which I keep extending into the woods as time permits, is well fenced to a height of six feet to keep goats in and dogs out. The goats can trot in under the barn to sleep in a big pile of loose mulch hay or to get out of the weather. At first I did the milking in the yard. Now there's a ramp in the barn floor beside the goat stall. I just let it down and the does come up to the milking stand, lured by an offer of grain.

Equipment

To hold hay, both in the stall and in the yard, I've found the most effective manger to be a three-foot length of four-foot-high welded steel fencing with approximately two-inch by four-inch openings between the wires. It is fastened to stall or

FEEDING SETUP

fencing in more or less a half-funnel shape. The openings are big enough so the animals can get their muzzles in, but small enough so they can't pull out great mouthfuls of hay, much of which would fall to the floor and be wasted. (As it is, dropped and "refused" hay acts as an automatic bedding replenishment, keeping the stall clean and dry on top, and by spring amounting to a good foot accumulation of the best garden mulch/fertilizer known to man.) The fencing is strong enough to withstand the constant butting, rubbing, and climbing on which it is subjected to, and won't get chewed up as fast as a wooden manger.

Our goat feeding setup

For graining I nailed a length of wooden storm gutter to one wall. Water is provided in stout buckets hanging beside the grain on a spring clip. Plastic buckets are best as they aren't hurt if the water freezes in winter. Access to grain trough and water is through a long door sawn in the stall wall. The door, kept horizontal by a chain at either end, also serves as the milking stand. I lure the "boss" milker onto the stand and into the stanchion by putting half her grain in a cup fastened to the wall. The other gets a half-ration while the first is being milked; then they switch positions. Like most animals goats are creatures of habit, and once trained to the routine, they go through it automatically.

Salt and trace minerals are provided by a salt block from the feed store which comes with its own wall-attached holder. If we lived near the ocean, I'd see if I could get the goats to eat kelp and get their trace minerals and salt naturally—and for free. But inland, we're stuck with the commercial product.

COMFORTS OF HOME

Goats can get smelly if not groomed frequently, but they are perfectly willing to do the job themselves if given the equipment. The bark of trees in the barn yard provides all the scratching post they need in summer. For the winter months I nailed bottle caps, serrated, bottom side out, on a three-foot-square section of the stall wall. This way they are able to get to just about all parts of their coat except perhaps the very top part along the spine. They adore being brushed there, however, especially at the base of the neck just forward of the shoulder.

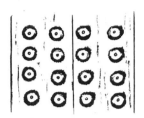

Bottle-cap   scratching post

The stall floor is oak, the walls double boarded to a height of five feet to keep drafts out. I cut a window six feet square, five feet up from the floor in the outer wall, installed hinged windows on the inside and bug screen outside. Another lesson learned the hard way: a mature goat standing on its hind feet can nibble well above the five-foot mark and they promptly tore the screen loose at the bottom. Now there is a stout board at the window bottom. To reduce drafts coming through the upper portions of the stall walls I nailed on double layers of old feed sacks. Needless to say, the lower sacks need an annual refastening or replenishment. To retain heat in winter, to let the milkers put their energy mainly into producing milk rather than working to keep warm, I also hung feed sacks from the ceiling in front of the top half of the stall opening into the barn. These are kept rolled up in summer, but let down in cold weather, they pretty much stop drafts and help retain enough natural animal body heat that the stall is comfortably warm even when it's well below zero outside.

Feeds

Goats are ruminants, which means they have an extra stomach or two that lets them stoke up on the tremendous amounts of plant matter necessary to sustain life. Like cattle, goats regurgitate newly browsed greenery—chewing their cud—and work it long enough to break down the plant fibers and add the enzymes needed to extract needed nutrients. This bulk, from hay or whatever, is not only desirable, it's essential. A goat fed nothing but corn wouldn't last long.

FEED
REQUIREMENTS

In the native state goats are browsers, enjoying new grass and some ground plants to be sure, though they prefer to feed on leaves and twigs. And not just on low-growing ones either. Goats can and will happily climb anything that will give them purchase; I've seen them twenty feet and more up into a huge rogue fir tree.

To start, our goats have free access to roughage. In early spring when the stored hay is showing its age I begin pulling new grass and snipping off budding twigs. As the season advances I switch from sickle and pruning shears to a scythe and limb loppers. A big armful of fresh grass goes into the mangers a couple of times a day and leafy branches are poked over the fence. (Goats go silly over the leaves from fruit trees, preferably from just-planted stock. A goat can prune them clean and girdle their trunks in less than a half-minute—another lesson learned the hard way.) Once they've nibbled all the apple prunings they get maple, which they relish. They also enjoy an occasional tonic of evergreen of just about any species, will nibble hickory and oak diffidently, but will flatly refuse to touch the sumac, aspen, or any other weed tree.

SUPPLEMENTS

With as much roughage as they can eat, goats need only a high-protein high-calcium supplement, the amount varying with the animal's age and status in the pregnancy/milking schedule. (The proteins and calcium that our children get from goat's milk have to come from somewhere. If I don't supply it the doe's milk factory will leach it from her own body and bones, and that's not good for anyone.) At first I toyed with the idea of mixing a complete and guaranteed organic supplement. But the goats wouldn't touch feed with bone meal in it; more than a dusting of calcium-rich limestone has the same effect. So, since our soil is notoriously low in calcium, I've made a compromise and feed half commercial dairy ration mixed with an equal amount of our organic feed of ground whole corn ears, limestone, and occasionally buckwheat. The commercial ration adds several protein sources such as soybeans that we just can't grow in New England, plus calcium disguised in molasses. The goats relish the stuff and do famously on it. Some day I'll find the time to talk with someone at the nearest Aggie college and see if we can't come up with a ration that I can produce completely and organically on our place. Might have to add Peruvian fish meal or something. I wish I could tell you I'd worked the formula out, but I haven't and won't till I can get some professional advice. Goats are too nice people to experiment on.

The whole lot of goats shares a good ration of supplement each noon, about a cup per animal. Both milkers and does in late pregnancy get an added cup each morning and evening at milking time.

Prunings for dessert.

Milking

One of those book-taught experts who says you can put goats in stanchion stalls suggests you learn how to milk by squeezing water out of a balloon. Which leads me to suspect a balloon is about as close to a teat as that "expert" has ever gotten. The milking process shouldn't start with a balloon, but with a pregnant doe about a month from her first freshening, i.e., a month from giving birth and coming into milk. One morning after a current milker or two have been stripped out and sent back into the stall or down the ramp into the yard, I will lure our newest mother-to-be up onto the milking bench. (She won't be totally unfamiliar with this procedure, having been through it during hoof trimming, which we'll get into further on.) There she will get a cup of supplement and I will start, very gently at first, working over the swelling udder and teats. She'll shy and kick at first but will gradually get used to it, so when she does freshen the milking will be a part of her daily routine.

The second part of the milking process is to be sure that no foreign objects at all get into the milk. Just one goat hair can ruin a batch. So the does get a periodic clipping; a small pair of barber's hand clippers is all you need to clip hair from the udder and shorten it along the belly. Then before each milking I rinse the udder and teats with a clean cloth soaked in warm water.

TECHNIQUE   To get the milk from the udder into the pail you should take a lesson from a young kid. It will approach its mother and butt the udder hard enough so it gets an occasional kick. But this is to get the milk to let down from the udder into the teat. You can do the same by grasping the teat and pumping it up into the udder—gently—several times. Then gently close your thumb and forefinger at the top of the teat, and sort of squish the milk out in stages by closing first the index, then ring finger, then the small finger against your palm. You never pull down on the teat except when the udder is empty, and you strip the teat clear by closing thumb and forefinger and running them all the way down. It's the easiest work rhythm and traditional to milk alternate sides and to sit while milking. I stand because the milking platform is at waist height and I do most milking with the right hand, usually having plenty to do with the left, such as keeping a kid from climbing up with its mother, catching the milk pail as the doe kicks it over, or rescuing one or another animal or tool from a three-year-old child helper. Worst mistake I ever made was showing off one time by squirting some milk directly into Sam's mouth. Now that's become a main event in our morning and evening chores, which are gotten through a lot more quickly without help, and speed is much to be desired on many a tired day.

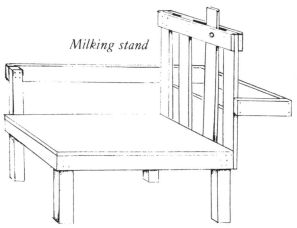

Milking stand

MILK QUALITY

The first squirt of milk from each side should be directed into a fine sieve. You can buy a stripping cup though an extra fine tea strainer will do; it does for us. If any lumps—small curds such as you find in milk that's gone sour—stay in the strainer, you likely have a case of mastitis on your hands. We'll get into treatment later, but for now just milk the udder completely dry, bury the milk, or feed it to the pig and be sure to sterilize the milk bucket. Never feed this ropy milk to humans or to very young goat kids.

KEEP IT CLEAN

To keep good milk good you have to keep it completely free of contamination. Even a current of air containing a strong odor can taint it. The experts (here they are again) tell you to put up a separate milk house with whitewashed walls and a hundred dollars worth of stainless steel buckets, strainers, and other dodads. That's fine (indeed required by law) for commercial dairies. I just use a double thickness of cheese cloth laid over a plastic milking bucket. This strains the milk at the same time it keeps out anything foreign, tainted air included. It simplifies the work too: the milking bucket goes right into the cooler, contrary to the expert's advice to milk into one stainless steel bucket, then pour the milk through a SS strainer into another SS bucket and have to wash three items instead of one. The plastic buckets, of special milk-house quality that have no taste or odor of their own, last about forever and the cheesecloth squares stand up to a good many washings before they get turned to dustcloths, cheese wrappings, or jellybags.

Care of the Milk Is Important

Goat milk must be chilled to about forty degrees as quickly as possible. In winter I just leave it, covered, beside the kitchen door. In summer it goes into a refrigerator or into the spring-house pit that's sunk into our cellar floor; the spring that is our water supply flows through it, and the constant flow of chill ground water acts as the best milk, watermelon, or cold-drink cooler there is.

The chilling is to preserve quality. If left warm too long the milk can develop a ''goaty'' off-taste. Once chilled it will stay good and unsoured for several days. It doesn't have to be homogenized, as the butterfat globules are much smaller than in cow's milk and it won't separate. This makes for easier digestion, I'm told, but it also makes for hard butter making. If you heat and chill goat's milk several times the fat globules will sort of get together and separate out into cream. Then after dashing it for a couple of hours you get butter and buttermilk. Altogether it's a lot of work, and the one time we tried it the end product *was* goaty tasting, so we've not tried a second time.

MILK HANDLING

Oh yes, you don't have to worry about pasteurizing goat's milk. Goats can't pass to humans the diseases that make pasteurization of cow's milk a legal requirement most places. You can pasteurize, of course, and it will keep the milk from souring a few days longer. A home pasteurizer costs about fifty dollars, though, and it's one more thing to wash. We just use the milk as is for drinking and cooking. Any left after a day or two goes into cheese. And if we're not in a cheese-making mood, there are always one or two litters of one critter or another around to lap up the excess.

NO NEED TO PASTEURIZE

I'll not go into cheese making. There are several good books on the subject out now and in any event you should buy rennet, an enzyme extracted from calves' stomach linings, to curdle the milk. Order it from Hansen's Laboratory, Inc., 9015 W. Maple St., Milwaukee, Wisconsin 53214, and ask for their excellent book of instructions. It was free last time I asked.

Breeding and Pregnancy

A doe's milk production is best the few months after kidding; then slowly it begins to fall off. So it's best to breed every ten to twelve months. A yearling doe can be safely bred at one year, though eighteen months is better for her overall condition. A doe kid can conceive at about six months of age, though such early kidding can harm the animal. I must admit, not all goat people agree on this one; some advocate early breeding, contending that by the time the young goat is carrying near-term young she's full-grown and in no danger.

DOE IN HEAT   Does come into heat every eighteen to twenty-one days pretty much year-round, though the normal breeding season is in the fall. They are amenable to a buck's attentions anywhere from fifteen minutes to several days during season, something that only a goat can tell. Anyone can tell when the general heat period commences, though. Some does fidget a lot, others try to mount other does, some flick their tails constantly, but they all "talk" about their problem in a most insistent manner. When I walk out into the barn one morning and a doe is peering over the stall door, bleating in a high, shrill tone, and when she keeps at it, even mumbling while she munches her grain, I know a heat is on us. If her vulva is somewhat swelled and inflamed with a slight discharge, the heat is confirmed.

MATING   After noting several succeeding heats you will know the doe's schedule and can arrange for breeding. It is usual to arrange the event, take the doe to the buck's place several days before a heat is to commence, and leave her for the buck's owner to breed. The heat may come on schedule but often is delayed from several days to a full three weeks as the doe reacts to a change of quarters. In any event, expect to pay from five to fifty dollars stud fee depending on the buck's qualifications, plus about a dollar a day for boarding and care.

PREGNANCY   A pregnant doe needs little special care until the last two months of her five-month gestation period. If she's milking, I dry her off late in the third month. This is a matter of milking her out completely, but only once a day for a week, then once every two days for another week or two. Toward the end of term does tend to get a bit lazy and spend more time lying down than is good for them. On our place at least one goat's last few weeks of pregnancy coincide with early fall. The weather is lovely, bugs are gone, and chores have dwindled somewhat, so we can spend a lot of time walking in the woods to go fishing or wading in the river that winds through the trees at the bottom of our hill. And we take the goats along. They never stray more than a few yards away, and they enjoy nibbling at strange plants and hopping around on the boulders. On many of these walks I'm prospecting for deer—searching out the game trails and deer browsing and resting places I'll need to know like the back of my own hand if I'm to have a chance bow-hunting for venison later in the year. And the goats' superior sense of smell and hearing are a big help. As we go through the deer woods their actions—ears perking, snorts and talking, trotting off briefly as if to join their distant cousins—tell me where the deer have been or the direction they are taking as they move ahead of us.

Birthing

A couple of days before a doe is due I haul the soiled bedding from the stall to the gardens, hose down well, and soak the wood with a disinfectant solution of Chlorox and water. After the stall has aired out I put on a new foot-deep layer of old hay bedding

Shovelling out the goat pen is never pleasant work, but there are a number of benefits. The garden gets fertilized, birthing takes place in a cleaner environment, and I get some more exercise.

and lead in the mother-to-be. Normally she will produce on the 150th day from breeding, usually in the small hours of the morning. It is advisable to be on hand on the odd chance she has trouble, but this happens very seldom.

KIDBIRTH   Some goat owners insist on playing midwife; I try to be around, but mainly I let nature take its course. There will be a liquid discharge that you might or might not notice, contractions that are plainly noticeable will begin, and in time the nose and front knees of a small goat will appear.

If contractions continue for more than a few minutes without noticeable progress it may mean trouble. A breach, where the kid's head is twisted back, blocking the birth canal, is very serious. The kid must be pushed back into the womb, the head located, and a normal posture produced. Sometimes legs are out of position and pushing back in and repositioning is advisable. In all of this procedure you must wear clean, preferably sterile gloves and since you are dealing with stretched and easily punctured membranes, any manipulation must be gentle and careful. Also the entire process must be carried out by touch. I'd advise you to call the vet; I would. Happily, we've never had birthing problems with the goats and if you get big, hardy stock the odds are you won't have any either.

Sometimes a kid will appear stuck, sort of quarter-born. It's just nature taking a rest. But if you want to help, don't pull the kid. Press against the doe's side as each contraction occurs. If the doe is standing—some do and others lie down—you'll want to catch the baby. In any event, remove any membranes covering nose and mouth, snip off the umbilical cord if it hasn't broken to three or four inches, and let the doe clean it up. This she'll do without any prompting. Then after a few minutes to catch its breath, the little kid will raise its rump into the air, get up on fore-knees and then stand up and begin wobbling around! It's an amazing thing to see.

The doe will produce from one to four kids, though a pair is most common. Afterbirth comes out in a blob shortly afterwards and the doe will want to eat it; I let her as nature intended. The only apparently harmful effect is that she will nibble disinterestedly at her food for three or four days, which is probably best for her system. She will have a considerable vaginal discharge, white ropy stuff followed by old blood, for several days after giving birth. It should have stopped in ten days to two weeks. If the discharge looks as though it contains pus and/or it smells foul, call the vet.

The young kids will not nurse automatically. Once the birthing is over and all is cleaned up, I wash the doe off with warm water only, and teach the new kids how to eat. It's a matter of holding their heads and milking out some of the first colostrum milk onto their muzzles, then opening their jaws with one hand and pushing the end of a teat with the other. It usually takes some time, particularly as the udder, just sprung to milk, will be quite swollen and tender.

NURSING

If you want to do as the goat milk dairies do and save all the milk for human consumption, you can milk out the doe and feed the kids from a bottle. Keep them from the doe and they'll never learn to suckle. This is pretty silly from our point of view. For one thing, there are one or two other does milking and

Kids are almost as cute as children; they are the cutest livestock babies, in my opinion. It's tough for any homesteader to remember always that these little fellows are ultimately milk, meat, hides, and a little bone fertilizer.

producing all we can use and more. The young kids do a better job of conditioning the teats and udder than a milker's hand, and the natural suckling sensation for some reason makes the doe's uterus contract faster, thus getting her into good shape again that much earlier. The kids' chasing her around give the mother needed exercise—to all appearances more than she wants. And besides, nature's way is a lot easier on the overworked homesteader. Why boil bottles when you don't have to?

GROWING UP   The little goats will grow at an alarming rate. After about a month the doe begins to resent their nursing, running away from them or engaging in butting contests as they try to suckle. They will have learned to drink from the bucket, nibble at hay and grain by then, so I separate the little family by stringing a length of four-foot fence wire across the back of either the stall or pen. They can nuzzle and converse with one another, but I will milk the doe and feed the kids their milk from a pan for the next four weeks. At two months of age they should be on normal goat ration and all the milk goes to us. After a week of not being nursed the doe will keep the kids away altogether, and after a week or so more they will stop trying.

Other Goat Chores

DISBUDDING   Some kids are born "polled," naturally hornless. Most are not, so you should make them that way before they are two weeks old. There are several ways to dehorn: one or two kinds of surgery, a hot iron brand, or using a caustic potash salve. Most experts recommend the caustic method, which suggests to me that they've never dehorned a kid and listened to it bleat piteously through hours of searing agony as the caustic eats down through the tender young skin and into its skull bone.

I use a disbudding iron which causes the kid only brief discomfort. First I check to see if the kid is naturally polled. There are two little spirals of hair on top of the skull. Under the hair is a small bump. If the skin moves easily over this bump, the kid won't grow horns. If the skin is tight, as is usually the case, it must be dehorned.

The iron—nothing but an oversized soldering iron—is heated as hot as it will get and....but wait a minute. This is one

Disbudding iron

procedure that oughtn't to be taught in a book. It's too easy to hold the iron on for too short a burn and a deformed horn will grow. It's just as easy to hold it on too long and I won't go into what that does to the little animal's brain. Nope, if you want to get goats, have the good person who sells you your first one, or have the buck's owner show you how to use the disbudding iron.

You might have the breeder show you how to castrate **CASTRATING** your young bucks too, though with the Elastrator there's no more to it than castrating a young pig. Just make sure the rubber band is up high against the belly when you let it go. Lacking an Elastrator you can use a knife if you have the courage. One person holds the young goat, the other uses a very sharp knife to make a one inch cut on each side of the scrotum. The whitish globule of testicle you pop out and cut loose. A dab of merthiolate, iodine, or purple horse medicine over each cut will complete the job. Lacking an Elastrator and the gumption to use a knife, you can do it the old fashioned way. Get about a yard of strong, fine thread or fishing line and make a noose in one end. Slip the noose over the scrotum up high against the belly, pull it taut, then wrap the line around, pulling tighter with each twist till you're sure the circulation is completely cut off. Then tie the line well several times. The effect is the same as an Elastrator.

And, even if you plan to slaughter your young bucks at the optimum age of three months, don't neglect to castrate and dehorn them. For one thing, I'll bet you put off the slaughtering time, maybe for good, with your first set of kids at least. And even at three months a young buck can service a doe, which can lead to some surprise pregnancies in the herd. The dehorning and desexing procedure may sound unpleasant and it is. But in time the black scab of the disbudding iron will come off, the tied-off scrotum will slough off, and you'll have yourself a nice young wether to train to pull your son's goat cart.

Hoof Trimming

Wild goats keep their hoofs filed down by clambering **HOOF GROWTH** around on rocks. Few dairy goats have such a chance and it wouldn't help if they did. The hoofs grow too fast to be worn down even if the goats are kept on a concrete floor. So hoofs must be trimmed, about once a month except for a doe in her last three months of pregnancy when a random kick could cause a miscarriage.

TRIMMING

The hoofs consist of a horny outer shell and a soft inner pad, called the "frog" for some reason. If neglected, the shell tends to grow beyond the frog and and will grow and fold over, the two halves enclosing the hoof like a deformed clam shell. The problem is particularly troublesome on the hind feet. Since an untrimmed hoof can lead to hoof rot, bone deformity, and resistance to needed exercise, trimming is a chore that shouldn't be neglected.

I put the goat on the milking platform and give it little dribbles of grain to keep it entertained. First I use a set of tinsnips to trim the extremely tough edge of shell down to the level of the frog. Then a sharp knife is used to trim both shell and frog as flat and even as I can till the frog begins to show pink. If the knife slips and cuts a bit too deep, drawing a drop of blood, it is nothing to worry about. Most hoof growth is at the toe end, and that is where dirt and manure will lodge, sometimes working up between shell and frog. This should be dug out, sometimes requiring you to cut out a bit of shell to get at it all. A goat, particularly a pregnant doe, has a lot of weight to carry on rather skinny legs and unless you keep the foundation—the hoofs—well cared for, you're asking for trouble.

SCOURS

Other routine care includes worming with a TBZ drench twice a year. External parasites, lice and the like, are no problem in our clean country environment. Some minor irritations do crop up. If an eye runs I bathe it gently with warm water with a bit of boric acid dissolved in it. Little sores on lips or under the tail are the goat version of a cold sore and go away of their own accord. Goats cough, sneeze, and break wind a good deal as a matter of course. A deep recurring cough, though, means a visit to the vet.

WORMING

Scours crop up occasionally. After a day in adult goats or immediately with very young kids, a mixture of warm water, molasses, and one of the people medicines containing Kaolin—plain chalk—will do the trick. The opposite, constipation, is fairly common just after a doe has delivered. I routinely give a new mother a drink of hot water and molasses just after delivery and twice a day till her droppings appear normal and in normal quantity.

ABSCESSES

Some goats are prone toward abscesses, lumps of infection just under the skin that grow to the size of a ping-pong ball. They are most common on the neck. At first I followed prescribed procedure, waited till the lump got big and soft, then put the goat in the milking stanchion, gave it some distracting grain, and lanced the boil along the bottom, squeezing out the glop inside. The problem was, the blasted things seemed to fill up again and again and repeated lancings made the animal skittish. Then if I got one cleaned out and healed, another would pop up next to it. Now I let nature take its course, isolating the goat and letting the thing drain unpleasantly but naturally. Then the animal gets a special ration including antibiotic-soaked niblets from the vet.

The most serious goat problems occur shortly after **MASTITIS** kidding time. If a newly-delivered doe acts listless, won't get up, has glazed eyes, or exhibits any other peculiar symptoms during the week or so after delivering, get her to the vet. An antibiotic shot will usually fix things. Check the teats and udder at least once each day. Sometimes knots will appear and you can sort of massage them out. If the milk appears ropy or lumpy or one side fails to fill as much as the other you have mastitis, which is a general term for several bacterial infections of the udder. See the vet again. He'll give you some big plastic syringes (for cattle) that you must use to infuse the infected udder with antibiotics. He'll tell you how. Time was you could get a three-medication mixture that would fix any mastitis bug. But for a reason that no one (except maybe a drug company lobbyist) understands, the government outlawed mixtures. Still, the great majority of mastitis is caused by one bug and the vet will know the best remedy to recommend. Needless to say, before administering an infusion you should milk the udder out. Then don't let the kids nurse so long as you are giving infusions and for at least two days afterwards.

If you catch mastitis in time you may nip it in the bud. More likely, though, the udder will be affected permanently to one degree or another, and will never give a full complement of milk.

If it sounds as though goats are a lot of trouble, they are. But what other medium-sized animal gives milk, meat, leather, free fertilizer for the gardens, and affection to boot? They provide an occasional added dividend too. For example, one hot fall day Louise and I packed baby Martha into the back-pack baby-carrier, cut Sam a hiking stick and, all of us stripped to essentials, took our four goats for a walk through the woods to the river. Coming into a favorite meadow we found a batch of uninvited city people spreading around beer cans, potato chip sacks and all for a picnic. From their point of view, there suddenly appeared from the trees a long-striding, long-haired bikini'd blonde with a baby strapped to her back, a bearded and booted type with a knife and fishing pole tucked in his underdrawers herding a two-year-old boy clad only in miniature hiking shoes, plus four deer! Their conversation stopped, jaws dropped slack. We nodded in a not-too-friendly fashion and the goats wandered, bleating, through the picnic. After the big wether ate their salad and left a reminder on the plastic ground cloth, we all disappeared into the woods. I don't know if they thought we were some wild bunch down from the mountain or what. But we weren't more than a few minutes away when we heard a car drive off. And when, after a refreshing swim we retraced our trail, I didn't mind picking up the litter they left behind.

Packing Produce In and Putting It By

10

One morning around the middle of September we awake to the high, clear sky of New England in the fall. The air will have a welcome bite to it and a walk through the gardens will show the top leaves of tomato plants beginning to wilt as the sun rises on them. By noon the effects of the year's first mild frost will show clearly; tomato tops blackened, morning glories drooping, a few leaves of the summer squash collapsed and darkening (and that's almost a relief as by then we've eaten and given away what seems a ton of it, and we never eat it from the freezer). But now the homestead takes on its final season of rush activity as we hurry to gather in the harvest and put it by for the coming winter months.

Of course, the harvest has been more or less with us since the fall-hatched pullets began laying and the old hens were sent to the stew pot in March, and we stretched out those first precocious asparagus spears in an April souffle. But from mid-September to the end of October comes slaughtering, freezing, pickling, canning, and all the other preserving work nearly full-time.

BLANCHING TIPS

There is no need for me to bore you with standard ways of freezing or canning fruits and vegetables. We do it the same as everybody: use only popping fresh produce, wash it well, blanch and chill it before quick-freezing, or cook it at prescribed times at prescribed temperatures in a sixteen-quart pressure canner. We have discovered a few things, though, that might not be in the back of every cookbook as are freezing and canning directions.

For one thing, timing of the blanching of vegetables

should begin from the second you put the fresh produce into your boiling water, not from the time the water reboils as is implied in some books I've read. Even if the water doesn't boil again, get the vegetables out and into ice water as soon as they've been in the hot water for the correct time. And for another, you will preserve more natural vitamins and minerals by steaming rather than boiling the produce. The good stuff is retained in the vegetable rather than being leached out into the blanching water. We steam everything but greens which would mat down and not heat evenly. In general it's best to cut vegetables in as small chunks as possible,

The woods around us is full of seedling apple trees, many of them happy to provide us with quarts of small, tart "crab" apples for jelly and wine.

then steam one third longer than you would blanch in boiling water. For example, Louise cuts broccoli into individual flowerlets and stem chunks into bite-sized pieces before steaming for five minutes, as compared to the three-minute boiling-water blanch.

Drying, or partial drying, to be more exact, was mentioned earlier. To store dried fruits, we put them in air-tight plastic bags that are kept in big old flour canisters. Out of the sun they will retain most color and all the special dried-fruit flavor. **DRYING** Louise uses the dried product to make fruit tarts, an occasional pie or good dessert of stewed dry fruit. She simmers it, two cups of water to each scant cup of dry fruit plus a little honey or sugar, then serves it or proceeds with her recipe.

Herbs are dried, sprigs are hung over the wood stove or are put in paper bags, hung upside down from the rafters of the ell

After the first light frost, Louise salvages squash and tomatoes that are still good though the parent plants are done for.

Our Preferred Ways To Preserve Produce

This isn't much more than a listing of our own preferences as to preserving methods. Not everyone will agree that green beans are better canned than frozen but you might be interested in knowing why we feel that way. So here is a list of the fruits and vegetables we grow and use plus the way or ways we put them up and why.

Vegetables

Asparagus: *There is some in the freezer that must be three years old. We think it loses too much texture in freezing. Now we gorge on it while it's fresh, and look forward to the next spring when it's not.*

Snap Beans: *Frozen beans retain good fresh taste but lose texture. We prefer them canned [boiled for five minutes, then in pints in the pressure canner for twenty minutes at ten pounds pressure and 240 degrees].*

Lima Beans: *If I ever figure out how to grow limas in our short seasons we'll freeze them by the bushel after a hot water blanch of from two to four minutes depending on size.*

Shell Beans: *Both bush and pole beans and yellow beans are let go to maturity late in the season. Shelled out, the dry beans are stored in glass jars in a dry place till needed.*

Beets: *Young greens treated as spinach; roots stored in the cold cellar in moist sawdust. Freezing or canning ruins texture and bleaches color out in our opinion.*

Broccoli: *Best freezing vegetable there is. Steam blanched for five minutes, chilled and frozen in pints.*

Cabbage: *Kraut. Heads stored upside down in the cold cellar. [Outer leaves removed, but roots left intact.]*

Carrots: *In sand or sawdust in the cold cellar.*

Cauliflower: *As broccoli.*

Corn: *Tried freezing on the cob and found extensive blanching ruined quality. Now we let ears become good and mature, but not real tough. Then husk, wash, blanch whole cobs in five minutes of steam. Then chill and remove kernels with a corn-cutter and freeze. Some is also let go to the near-dry dent stage, where there is a little indentation in the top of most varieties. Shelled from the cob, stored covered so it won't dry out too much, it's cooked in oil or butter till it swells and we have parched corn to go with the apple cider on a cold winter evening.*

Greens: *Spinach, kale, beet greens are all blanched in boiling water for two minutes, chilled and frozen. Some kale wintered over in the ground under a cold frame arrangement.*

Okra: *Okra, which we grow in small quantities for gumbos, is blanched for three minutes and frozen.*

Onions: *Stored in hanging sacks in cold cellar.*

Parsnips: *Left in the ground under heavy mulch all winter and dug as needed.*

Potatoes: *Stored in maple leaves in cold cellar.*

Peas: *Frozen after two minutes blanching in steam. Good freezers. Some let dry.*

Peppers: *Hot peppers dried and strung in the kitchen.*

Green bell peppers *will keep for several weeks in refrigeration. To freeze we fry slices in hot oil for a couple of minutes, cool and freeze in small quantities in plastic sacks.*

Pumpkin: *Split, seeds removed and retained, roasted in the oven till tender, then the pulp is scooped out and frozen in pint or one-pie containers. Seeds are removed from the stringy stuff and roasted in peanut oil in a medium-warm oven till crisp and delectable.*

Squash: *Winter varieties are let sit in the fall sun to cure till frost. Then, stored off in an odd corner of a warm closet they keep all winter. Seeds used like pumpkin seeds. Summer squash is enjoyed when we have it fresh. Turns to mush frozen. A great deal of it goes into the vegetable soups we cook up during the growing season and freeze for winter lunches and suppers.*

Tomatoes: *Used to can them. After a tip from Ruth Stout, Louise now just squeezes out the pulp and freezes them whole, skins and all, Easier, quicker and just as good in spaghetti as the canned product.*

Fruits

Apples: *Some are steam-blanched briefly and dried. Most red apples are kept whole in the cellar bins. White apples are canned as apple sauce or cut up as old-fashioned "applesass"—packed in a crock that's filled with boiling cider, then stored down the cellar.*

Apricots: *Seeded halves are dried after five minutes steam-blanching.*

Blackberries/Raspberries: *Those that aren't eaten fresh go into preserves.*

Blueberries: *Half-pints covered with a half cup of sugar are let sit till juice develops, then are frozen.*

Cherries: *Sour variety only, treated same as blueberries.*

Cranberries: *Can't say we have our own bog, but when the crop is new we buy a half dozen boxes and freeze them as they are till they're wanted to go along with a big chicken dinner.*

Grapes: *Wine and jelly.*

Peaches: *Dipped in hot water till the skins come off readily, pitted and sliced and frozen with a sprinkling of sugar in half pint freezer containers. A bit of orange juice provides vitamin C to prevent discoloration.*

Pears: *Peeled, cut in slices, heated for one minute in a syrup made of three cups of sugar dissolved in four cups of water; with a bit of orange juice added, frozen in half pint containers. Pears are picked when still hard and for fresh eating they will keep in refrigeration for many weeks.*

Plums: *Tree-ripened fruit halves are blanched in steam for about five minutes and dried.*

Rhubarb: *Cooked for a minute in the "three-to-four syrup" used for pears, frozen in half pints with or without strawberries, for pies.*

Strawberries: *Jam and, either whole or sliced, packed in half pints along with some orange juice and ¼ cup of sugar per quart and frozen.*

where fall heat dries them in just a few days. Then the leaves or flowers are stripped off and stored in air-tight spice jars.

An old-timey dried food is snap beans, picked at their prime and strung with needles on thread. It's called "leather britches" and in our opinion both flavor and texture match the name. The beans look good hanging from the rafters, though. The remnants of our one stint at bean stringing have become a permanent ornament.

From the little herb garden to her right, Louise annually harvests a year's supply of sage, mint, tarragon, basil, and several other herbs.

The Cold Cellar

We are fortunate in having a spring that flows under the concrete floor of the cellar. It keeps humidity up and normal cellar temperature at about fifty degrees year round. However, by putting a good layer of insulation on the ceiling and installing an insulated wall, we've managed to create a cold-storage area at one end of the cellar where temperatures of just about freezing can be maintained from late fall through the spring. Since this is the best temperature for bin or hanging storage of most fruits and vegetables, the cellar serves well as our winter produce bin. The cool temperature is also fine for smoked and dried meats and most winters find hams, slabs of bacon, and smoked sausages hanging from the overhead.

I installed one of those indoor/outdoor thermometers in the kitchen with the sensor on a wire that is normally run outdoors (when the thermometer is installed in the house) running down into the cold cellar. Thus the inside reading tells us the house temperature and the outside reading the temperature in the cold cellar. If the cellar threatens to go much below thirty-one degrees I have to go down and open the door in the insulated wall a crack. If things get a bit too warm there is a window that lets cold outside air into the cellar.

Along one wall of the cold cellar are rough wood shelves for canned foods, jellies, wines and the like. Along the other are deep wooden bins and a rough wood closet with shelves that were used for food storage by the original owners of the house, before central heating was invented. One bin contains sand or sawdust to a depth of two feet. (It's important to obtain hardwood sawdust, by the way. Any pitch pine sawdust would ruin many vegetables.) As they are harvested, we pack in carrots, beets, and winter radishes. Bins are thoroughly moistened in the fall and kept that way with an occasional spraying during the winter. Sawdust, peat moss, or other organic matter seems to be far better for keeping roots than the alternative, sand. Besides being heavy as all get out, sand dries faster than any organic storage medium.

SHELVES, BINS, AND SAWDUST

The other big bin gets packed with apples packed in layers so no two touch, in newly-fallen maple leaves. The leaves are kept as loose as possible so that air can circulate slowly through the fruit, vegetables, and packing. Apples and potatoes like a moderate amount of moisture so the leaves are sprayed just a bit if they are well dried at storage time. Then from time to time we soak down a good layer of sawdust that's spread under the storage bins. As the water evaporates it keeps the air moderately humid.

Onions like it cool but dry, so they are hung near the ceiling on the wall opposite the bins. The years we get them in early with a good length of stem attached they are braided the old-fashioned way. When harvest is late and the tops are gone I hang them in string mesh onion bags.

Cabbages were hung upside down from the cold cellar overhead our first year, till we realized the house was beginning to smell like an improperly maintained compost pile. Now we pull the plants, strip off most outer leaves, and pack the heads, roots, and all upside down in cardboard boxes packed with maple leaves.

Then the boxes are piled on the cellar steps and both the slanted ground-level outside doors and cold-proof inner doors are closed. Enough cold comes down from above and enough heat radiates out from the furnace and heat ducts to keep the vegetables happy through the winter.

The Pickle Shelf

For my money the stars of the canning shelves are the pickles in their various shades of green, yellow, and red. Louise opts for the jellies, but then she's the jelly maker. I make the pickles and, along with my West Virginia brother-in-law, eat most of them. The problem most folks have making pickles is that they try to get too complicated, or at least that's our experience. One year I believed a recipe book and made half the cucumber crop into a fancy mustard pickle that required making a sauce and adding fancy spices. The cook book writer raved about it. We though it was awful—sauce gummy and sour, vegetables flabby and unappetizing looking. The pigs even refused to eat it.

Now we stick pretty much to three basic recipes, making minor changes in spices or herb ingredients as the notion hits. All three are simplicity itself, one sour, one tart, the third sweet. You can put up any vegetable in either juice, though the sour one is a basic dill pickle juice, the tart one for relishes, and the sweet one for bread and butter pickles. The sour pickle juice consists of one quart of mild apple cider vinegar, two quarts of water, and a cup of

THE SOUR PICKLE coarse pickling salt (flaked Kosher or granular sea salt). This mixture is kept in a gallon jug and is replenished whenever necessary. Into it goes just about any vegetable that we have excess of on any given day: green beans, broccoli heads, young zucchini squash, small and firm green tomatoes, and of course, cucumbers. I'll sterilize in boiling water enough of our old-fashioned glass-topped canning jars to hold the produce, then bring a pan of pickling brine to a boil. Into the bottom of each jar goes a fresh grape leaf—to give the produce a healthy pickle-green color—a head of fresh dill from Louise's herb garden, and powdered alum, 1/8 teaspoon for a pint jar and 1/4 teaspoon per quart. For large cukes we want as dills, either whole or slices, I add a big cut clove of garlic. For dilly beans I add a small red hot pepper. Green tomatoes receive a hot pepper and a small pinch of celery seed. A small pinch of sugar is added to each jar, not to sweeten the brine, but just enough to soften the bite of the vinegar a bit. Finally, the just-picked vegetables are scrubbed well, packed into the jars, and I fill to overflowing with boiling brine and seal.

A quart jug of the sour pickle containing dill, a tablespoon of mixed pickling spices, and a couple of pinches of mustard seed are kept in the refrigerator. We use it to produce ''half-sour'' pickles; after a day or two of soaking in the brine a medium-sized cucumber takes on a delightful tang. This same refrigerator pickle is used to fancy up hard-boiled eggs, little sausages that have been simmered on the stove for a half-hour, and the carrots, onions, sweet peppers, and cauliflower that needs a week's brining to give an authentic old-world flavor to a big antipasto.

THE TART PICKLE

The second, tart, formula combines a quart and a half of cider vinegar, a pint of water, a half-cup of pickling salt, and two cups of sugar. Into this goes a tablespoon of both celery and mustard seed and a few black peppercorns. With this tangy-sweet pickle we make all manner of vegetable relishes. My favorite is New England corn relish which gets put up by the shelfful when the sweet corn is coming on in abundance. I wait till the last of a planting has gotten a bit old, the kernels a bit tough for best eating, but not dry to the dent stage. For about a dozen pints of corn relish we cut the kernels from a baker's dozen big ears of corn, chop coarsely a head of cabbage, a half-dozen medium-sized onions, two or three sweet peppers, and from three to six red hot peppers, depending on size. Sometimes I add in green beans or cauliflower, also cut in small chunks.

The vegetable mix is put in a bowl and covered with part of the two-quart pickling juice. The rest goes into a big pot along with a couple of pinches of turmeric, a spice used mainly for its rich, golden color. I bring the pot to a rolling boil, then pour in cornstarch mixed with a little water. The starch cooks immediately, and I keep adding till the juice is thick enough so the bubbles pop audibly when they come to the surface. Then the vegetables and the rest of the pickle are added, all is mixed, simmered for a half-hour, then packed in sterilized jars.

VARIATIONS ON TART

The variations possible with this pickle are endless. A mix of two parts chopped green or red peppers to one part chopped onions make pepper relish. Replace the corn with carrots and you have a vegetable relish. Cut up apples, peaches, or just about any fruit you please, add in a good dose of raisins, a mashed garlic clove, powdered ginger and/or curry powder and/or the curry-like

spice fenugreek to taste and you have a chutney. Thin slices of a seeded lemon add more character to a chutney and the thickener should be flour, not corn starch. Cooking time for any of these will depend on ingredients, the idea being to get the fruit or vegetable tender, but not over-cooked and mushy.

THE SWEET PICKLE   The final, sweet, pickle recipe will turn out wonderful bread and butter pickles, sweet picnic chips, and gherkins. You start with equal amounts of mild cider vinegar and a mild but tasty partially refined sugar (commercial light brown sugar is OK). For about a dozen pints of pickles I mix a quart of vinegar, four cups of sugar, a teaspoon of turmeric, a half-teaspoon of ground cloves, a teaspoon of celery seed, and two tablespoons of mustard seed. This mixture is brought to a good boil, vegetables are added and heated

Harvesting the honey is simply a matter of removing the filled super from the hive, as Bill Corey has done [opposite]. With the super on the kitchen table, Louise cuts the honeycomb from the frames, deposits the pieces in a pot, and slides the pot into a moderately heated oven. The wax is melted and the honey sterilized by the heat. Once cooled, the wax, along with chunks of pollen, dead bees, and the like, comes right off. Both honey and wax are strained, the honey bottled and stored, and the wax either sold or used to make some of the world's best-burning candles.

to not-quite boiling and packed in sterilized jars. The traditional bread and butter pickle consists of medium-sized cucumbers, onions, and green and/or red peppers all sliced very thin. You can vary the vegetable proportions to suit yourself; I generally add a cup of onions and one pepper to each five or six cups of cukes. It's best to salt the water out of the vegetables before heating them. I mix a quarter-cup of salt with each six to eight cups of vegetable and let it sit for three hours. The vegetables are drained and rinsed once in cold water before cooking. The same procedure is followed in making sweet cucumber chips, except that onions and peppers are omitted and the cucumbers are cut in quarter-inch slices. To make sweet gherkins you need cucumbers only, preferably small ones of roughly the same size, and before processing they should be soaked for twenty-four hours in a brine mixed from one part salt and nine parts water. All these pickles should be let sit for a month before they are opened, and we find that they lose quality after more than a year's storage.

"Krauting"

Not too many people make their own sauerkraut any more, and a good many who have given it a try quit after the first attempt. The problem, which seems to be conveniently skipped by many book writers, is odor. If a batch of kraut goes bad it can put out a smell that will clear your head real quick. So don't try making it in the kitchen. The downwind corner of the barn is more like it. And stick *strictly* to the salt/produce ratios.

Krauting is a process of controlled fermentation in a weak salt solution. Naturally occurring bacteria turn vegetable sugars to lactic acid and then the acid and salt keep decay bacteria and yeasts from growing, so krauts will keep for some time. The process retains much of a vegetable's Vitamin C and to many a New England pioneer family the kraut tub was the difference between a healthy winter and scurvy.

You can kraut just about any firm-fleshed vegetable, though members of the cabbage family are most commonly used. We kraut cabbage, of course, as well as rutabagas and their tops, and the big winter radishes with happy results. Attempts with carrots and green tomatoes were disappointing. About the only vegetable we kraut with regularity is cabbage; our late Savoy cabbage is especially good.

We usually make ten pound batches. Three or four cabbages are brought down from the garden, outer leaves are given to the goats and rabbits, and the heads are let sit overnight to wilt a bit. Next morning they are weighed, quartered, and shredded into a straight-sided cooking pot that holds about four gallons. For a ten-pound batch I measure out a careful half-cup of pickling salt which is mixed thoroughly—very thoroughly—by hand into the kraut. When the cabbage has gotten limp enough so that it won't break, I tamp it firmly into the bottom of the crock with a potato masher. There should be enough water drawn out by the salt to cover the cabbage. A clear plastic bag filled with water goes on top to hold the vegetable under the brine; several sticks slipped in between the crock side and the bag will let the fermentation gasses get out more easily. A cloth towel is fastened over the top with an elastic band to let gasses escape and keep bugs out.

The kraut should be kept at warm room temperature for a couple of weeks. We check it daily to be sure the brine is high enough to cover the vegetable. If it falls I top up a cup of water having a tablespoon of salt dissolved in it. Every day the accumulation of scum that bubbles up is scooped off and after working for ten days to two weeks the bubbling stops and the cabbage is kraut.

Time was I would lug the container down into the cellar where the kraut would keep all winter and then some, so long as the brine covering was kept up. But since we ran into an inheritance of a hundred or so extra canning jars we put it up in glass. The kraut is packed tight in sterilized jars, the brine with a cup or two of water added is brought to a boil and poured in to overflowing, and the jars are sealed.

A delightful variation is to sweeten the kraut. I like to make up a kraut juice version of the second pickle discussed above, the tart one. A cup of sugar goes into each quart of kraut juice along with a teaspoon or two of mustard seed, a sprinkling of celery seed, and shreds of hot red pepper. This sweet kraut is put up the same as the sour version. It's more or less a preserved cole slaw and is best served cold on a leaf of lettuce during a hot weather meal.

The Winter Garden

KALE, PARSNIPS AND LETTUCE

The kale and parsnips shielded from the weather by hay bales and boards are one half of our winter garden; the other half is lettuce kept over winter in the cold frame. First I move the frame over to the house, backed up against a cellar window. Then into the soil I transplant a couple of dozen plants of the super-hardy lettuce variety **Oak Leaf** that has been started in early fall. The ground next to the house is relatively warm and I can open the cellar window a bit in bad cold snaps.

This lettuce pretty well stops growing as soon as the days shorten in October, but it withstands the cold remarkably well. If kept from freezing completely solid it retains its crispness and provides us small but greatly appreciated salads well into the winter.

Winter Feed for the Animals

When it comes to putting by grain and forage for the livestock I really am thankful that we live in the 19 rather than the 17 or 1870s. For one thing, there's always the feed store if the home-grown supply runs out. And instead of hand scything,

HAY raking, shocking, and hauling an acre or more of hay, I just drive the truck up the hill where a good neighbor keeps the family farm in production (and his father's and grandfather's spirits more or less contented) by haying his fields of natural meadow grasses every summer. He does it more for the smell of new-cut hay and the pleasures of good, hard work, I suspect, than for the little he charges for a ton. Most of the cash the hay brings in goes into keeping the old equipment going.

GRAIN But grain is a problem. For one thing, it is expensive now and going up. For another, commercially grown grains are about the most chemicalized, bug-sprayed crops there are. So we grow our own field corn and small grains as detailed in an earlier chapter. Of course, they must be harvested and stored for winter use, a procedure that is seldom the same from year to year due to weather. A wet, cold spring delays planting so the corn fails to ripen and mature. A very dry summer will cut the grain crop dramatically while a very cloudy and wet late summer will produce lovely stalks but little grain. So what and how we harvest varies.

Preferred is a perfect early planting and a good hot summer with plenty of rain. Then we bring in foot-long ears of corn well dried for grinding and storage sometimes before mid-September frost. We make a neighborhood party out of the harvest, with a feast of sweet corn in the evening to top off the work. I let the teenager currently learning to drive man the pickup and as many people as are willing go along the rows breaking heads and ears and throwing them in the truck bed.

The half-acre or so we plant to grain each year will produce at best fifty bushels of corn. Figuring the weight at fifty-five pounds per bushel, that would give us about twenty-seven

The best way to fatten up the goats against the winter cold is to turn them into the corn field to harvest the runty ears, stalks, and leaves.

hundred-pound sacks of grain. If you believe commonly accepted ratios of acreage to food production, in a year's time that could get us one Choice-grade yearling beef steer, four 220-pound hogs, 3,000 quarts of milk, or 800 dozen eggs. What we need, basically is grain to provide perhaps one-quarter of the feed for the equivalent of two hogs, one fifth of a half-acre's milk production to get us some 600 quarts of goat's milk, and about one-seventh of the feed needed to produce 800 dozen eggs, as we use only about 120 dozen in a year, or four per day. What this boils down to is that a little more than half of the field grain goes into pork, milk, and eggs. The rest of it goes into growing young goats to be future milkers or future meats, and into chickens and rabbits for meat.

FEED HUSK, COB AND ALL   We don't shell the corn or strip off the sunflower seeds. It's a waste of time, and I figure that there are some nutrients and a whole lot of good roughage in the cobs, husks, and flower heads. So, once we get a truckload of corn back to the barn, we crank up the shredder/grinder and run a few ears of corn through. If it is dry enough the ears will literally explode into chicken-nibblums. Then we take a break and go through the dried-out sweet corn for culls and to pick up the drying gourds, pumpkins, and squash, which usually amount to five or more bushels. Then the grinding begins. The big shredder has powered wheels and I just run it up onto the truck bed, hang a feed sack over the discharge chute, and the kids begin feeding in grain, steady sprinklings of dolomitic limestone, plus heads of whatever small grain we may have grown that year. Also added are gourds, pumpkins, and squash if they are dry enough.

CASH SAVED   All-in-all the harvesting and grinding is a tough day's work for Louise and me, a Sam and Martha sitter, and three or four teenagers, plus whatever adult help happens along. But for the labor we have some fifty 100-pound sacks full of animal feed that would cost $500, transportation not included, from the feed store. But more important, it's a grain produced without a drop of poison or artificial fertilizer—cheap at twice the price if you could buy it. Which you can't.

When the Weather Doesn't Cooperate

SILAGE   Many years our growing conditions are less than ideal and the grains do not mature enough to be stored dry. Then we make silage. The ensiling process is nothing but a way of krauting field crops, stalks and all, except that in the absence of salt brine, the fermentation process is more complete than with a kraut. The end product is pretty rancid smelling, a brown color and much lower in protein than grains. But the livestock adore the stuff and do well on it, though the chickens must spend the winter on bought feed if we want to have our morning eggs.

Cutting silage takes the same crew as gathering mature field crops. The difference is that we are pulling whole plants. There is an L-shaped gadget called a corn knife you can use to cut stalks at ground level, or a machete would serve. We figure that the plants' roots contain a good deal of the nutrients that didn't have time to get turned into finished grain, so we pull the whole plant, knock off as much soil as we can, and grind it up roots and all. We figure we're doing the livestock more of a favor than harm in letting them ingest a bit of our good organic soil.

The stalks are piled into the truck, then run through the shredder/grinder set so the shreds are an inch or so in length. The raw silage, like the dry grain, is run into the new-style plastic weave feed sacks, but it is packed in as tight as we can make it. The reason is that proper ensiling takes place only in the absence of air.

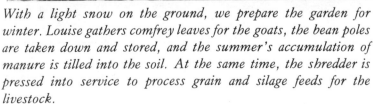

With a light snow on the ground, we prepare the garden for winter. Louise gathers comfrey leaves for the goats, the bean poles are taken down and stored, and the summer's accumulation of manure is tilled into the soil. At the same time, the shredder is pressed into service to process grain and silage feeds for the livestock.

Introduce oxygen and the silage would rapidly turn to compost. To keep the minimum of air from getting in we roll the open tops of the sacks down tight, then pack sacks as close together as possible on a big plastic sheet that's then folded over the stack and held down with hay bales. The stack goes on the dirt floor of the lower barn level, the side fenced off from the goats that would use the stack as an edible play mountain if they could get at it. You daren't keep silage in a barn loft, by the way. It drains all the time, a pungent and slightly intoxicating brew that would soak into the floorboards to stay. Better to let it drain off into the soil.

The years we get a good, dry grain yield, storage is in the loft. Rats, mice, and chipmunks would have a field day if we just piled up the sacks. So I creosoted floor and walls of a loft corner, patched holes with aluminum flashing and fenced the corner off with five-foot-high aluminum window screening hung as floppily as possible from the rafters. No creature will climb more than a foot up a real floppy fence.

Winemaking

The final phase of our harvest is making the wine, a process that begins when the first grapes or berries ripen and which extends well into the winter months. There is a whole industry that's popped up over home booze-making with shops selling expensive equipment, wine brewing sets you can buy for Christmas presents, and any number of books. Most are by Englishmen and give such mysterious directions as "Combine a half litre Demerrara sugar with a twidgin of Treacle, coddle with 1¼ Imperial gal. coltsfoot spirit or eau'd Whortleberry..." We've tried most of them and gotten some retchingly bad results. It may be that neither Louise nor I was made to follow directions very well, but we've had the best wine-making results by doing our own experimenting. Recipes you can get from any number of

OUR TECHNIQUE books. But here's how we do it "by guess and by grape."

First, we mix juice that fills the following requirements:

1. It's a good light fluid, not thick and heavy or so thin the flavor is weak.

2. It contains sugars (there are several natural kinds, not all tasting sweet) enough to make a pleasant drink.

3. It contains acid (several kinds of those too) enough to offset the sweet, which by itself in insipid.

4. It contains a bit of bitter, enough to give "character" to the taste, but not enough to be unpleasant. Then we add yeast and let the juice ferment naturally, bottle and age. The results...wine.

The idea of a wine, I suspect, is to provide a flavor that is complete, one that stimulates all taste buds (which sense sweet, sour, and bitter) in a balanced way, if that makes any sense. The alcohol content is secondary...to us, at least. The flavor should also include aromatic elements that please the sense of smell, which is where most of the sense of "taste" really resides. And it makes sense to start out with a juice that also satisfies these requirements. I think of freshly squeezed orange juice as such a perfect combination. Many carbonated soft drinks are formulated to offer a balanced tongue taste, though the smelling part is nothing but fizz water.

STARTING JUICE

In putting together the starting juice, we try to balance each of the individual flavor components as they occur in fresh orange juice (not trying to match the orange *flavor* at all, but building a flavor that is as complete as orange juice). It isn't hard, you just have to concentrate on sweet, sour, and bitter separately as you taste. The smell part you can't do much about except keep your fingers crossed.

First step is to prepare the raw juice for the preliminary fermentation. There are three ways, depending on raw ingredients. Grapes are removed from the stem (about thirteen pounds for each gallon of wine) and put in a plastic tub big enough to hold twice the amount of juice we'll be working with. A big plastic trash barrel is what we use, and it gets washed with salt and baking soda only. Soap or detergent will leave a film that can ruin wines. With hands well washed and rinsed, I crush the grapes with a potato masher till all are broken.

Fruits other than grapes must be smashed as much as necessary to break up the cells that hold the juice. Strawberries need only a brief squashing while cranberries have to be boiled. When we make the basic juice for flower or herb wines, we add equal amounts of herb leaves or flower petals (no stems, caps, or cores, but *petals only*, whether rose or dandelion) and boiling water and stir briefly.

SWEET AND SOUR

This beginning pulp is let settle for a few minutes till we can get a half-cup or so of pure juice. This Louise and I both taste, adding in ingredients till we have a juice that pleases at least one of us. To add sweetness, we stir in sugar. Honey can be used, but to my mind sugar is better. The delicate taste of honey is easily buried by most fruits, and, more important, honey ferments very slowly, markedly slower than sugar. To add sour, we add fresh, canned, or bottled orange or lemon juice. For both sour and bitter, we add grapefruit, cranberry, or sumac juice. To add pure bitter, we add strong tea or a brew made of boiled acorns (and that *is* bitter). Of course, each addition is measured and recorded; so many half-teaspoons of sugar, so many drops of this or that. Using, say, a half-cup of juice, we simply multiply the amounts added by the number of half-cups of juice we have, and then dump in the appropriate amounts of additives.

One additive that should not be overlooked is plain water. Often a juice is too sour or bitter—too strongly flavored in either category—and addition of sugar alone won't fix it. This is always the case with our wild grapes. Often we must add almost half again the amount of water to cut the acid content to a palatable level. The resulting wine is very light in ''body'' and is a lovely bright color.

GRAPES   Anything poured into *grape* pulp is boiled first for ten minutes and allowed to cool, since we ferment grape wine with the yeasts (microscopic plants that turn sugar to alcohol) which naturally grow in the white, dusty ''bloom'' that covers the fruit. Hot water would kill them. Once doctored, the vat of grape pulp is covered with a plastic sheet tied loosely and set in a warmish corner of the house (about seventy degrees) to undergo preliminary fermentation.

To make wine from fruit juices *other than grape* the sugar, water, or other doctoring ingredients go in boiling hot. This doesn't harm the fruit, but does kill off most of the germs and ''wild'' yeasts that might spoil the wine. It isn't a good idea to boil most fruits as the flavor is changed and aroma killed, though there are exceptions: cranberries as mentioned earlier, rose hips, crabapples, and most hard fruits. Of course, juices from flowers and herbs were made by soaking the petals or leaves in boiling water, so they are already perfectly sterile.

FERMENTATION   If you want, you can use commercial fermenting yeasts; they are sold at any wine supply outlet and give uniform results. Do *not* use baking yeast or your wine will be cloudy and smell and taste like under-baked bread. And the brewer's yeast sold as a food supplement won't work at all. We have used commercial fermenting yeast from time to time but prefer to grow our own from grapes. Here's how we do it. First, a good bunch of grapes having a strong ''bloom'' is picked over. All stems, bits of leaf, spider webs, etc. are removed. The grapes are swished around, squashed, then let sit in a quart jar half-full of warm water that has been boiled with an equal amount of sugar for ten minutes, then let cool. We cover the jar with a piece of plastic, held securely but not tightly with a rubber band. Kept in a warm place, the yeast on the grapes will begin to grow. In a few days there will be a good bubbling yeast culture going that fills the jar with froth and keeps the plastic cover stretched tight. When the yeast activity has reached a peak and before it has begun to subside, it's time to add it to the juice or refrigerate it for future use. It will keep for months.

Commercial fruit press

From here on in all wines are treated pretty much the same, whether grape, other fruit, or flower based. After the yeast is added and stirred well into the pulp, the tub is let sit till the plastic film top doesn't puff back up from a poke. This takes from two days to a week, sometimes more, and signifies the end of the first fermentation. Next the juice is poured off the pulp and what remains, be it grape skins or flower petals, is squeezed. Petals can be wrung out in a big jelly bag. Grapes must be pressed either in small lots in a jelly bag or in a fruit press. We got a small one from one of the big mail order houses a few years ago for less than ten dollars. Now the selfsame thing costs over three times the amount. In my opinion it was overpriced at ten dollars and is a flat gyp at over thirty. I think if I had it to do over again, I'd build a gravity press similar to a giant cheese press. You support a large tin can, a three-pound coffee can or larger with the top out and the

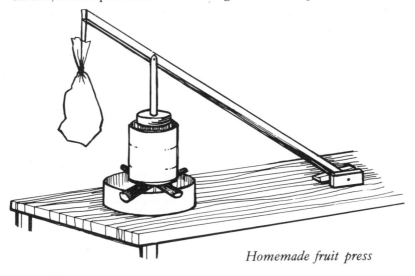

Homemade fruit press

bottom punched full of holes, over a pan with crossed sticks. Then attach a follower, a wood circle just a bit smaller than the can opening, to a long hinged lever and squash the pulp with a weight at the lever's unattached end. The pulp should be in a stout fabric sack or it will ooze out the holes. If the pressing area is cool and you arrange to cover the press to keep bugs and air-borne yeasts out, you can let it sit overnight to get all the juice.

First Year Aging

Once the juice is pressed from the pulp, we ladle out a cup or two, bring it to a simmer, and dissolve in it at least a half-pound of refined sugar for each gallon of juice. (Some winemakers will tell you this isn't enough sugar, that you should use as much as two pounds for each gallon. If the fruit you are using is not naturally sweet, you may need more sugar. Our preference is to make ''dry'' wines and add additional sugar to those batches we decide should be sweet wines, as I'll explain presently.) The juice and sugar are boiled for ten minutes to melt and sterilize the sugar; it is let cool and added to the juice. Then the whole thing is stashed in the cellar and forgotten for a year. The wine-supply stores sell white oak casks to age wine in. Booze makers have cut down so many white oak trees the species is actually endangered, so casks cost a minor prince's ransom. Besides, they are coated inside with paraffin, so the wood is only for show. I suppose that some very strong red wines would benefit from soaking up tannin or whatever from an untreated wooden cask, but we'll leave that to winemakers with more free cash.

AGING　　　We store our wines in three-gallon white plastic milk pails. The plastic is nontoxic and gives off no taste or odor, which is more than you can say of the cheaper plastic buckets you can buy in any dime store. Glass carboys would be better, if only because they are so easily sealed—a 10 cent balloon will do it. I'm told that spring water bottlers will sell carboys, but we haven't found one around here that will. The squeezed-out and sweetened juice goes into the buckets, a stout plastic sheet is tied on with a good quality elastic—tight, but not so tight that fermentation gasses won't be able to escape—and the lot is put on a shelf in the semiheated part of the basement. There it sits at an average temperature of fifty-five degrees for a year.

What happens when juice turns to wine is a mystery to everyone, including the wine chemists. All you can do is wait, hope, and make sure the fermenting juice isn't disturbed. It will begin to "work," the yeast generating carbon dioxide gas as it slowly turns sugars to alcohol. The gas will distend the plastic cover like a balloon. The excess will escape, but enough should remain, so that no foreign substances, including air, can get in and ruin the wine.

After a year we sneak a long, thin, sterilized plastic tube into the vats and siphon out a small bit of the juice. Put one end of the tube into the juice, let it fall in a loop that droops below the vat bottom. Raise the other end higher than the vat top and suck on it. Liquid will fill the loop and as long as you can keep the loop bottom lower than the vat bottom, you can siphon; just put the free end into a container and lower it. To stop the flow, raise the free end.

TASTING

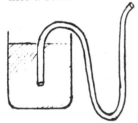 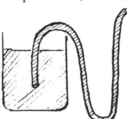 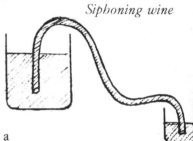

Siphoning wine

You never know what to expect with a wine. Sometimes a batch has all turned to vinegar while another hasn't even fermented enough to lose the sweetness. That is when we are glad we make wine in three-gallon batches, rather than really big lots. Almost always, though, especially if the covering has remained tight, you will have wine. It will be "dry," with all or most of the sugar turned to alcohol. We make a decision at this point whether to bottle the wine and use it as a dry dinner-type wine or to sweeten it. This is a matter of personal preference. If we decide to bottle, we wash in detergent, not soap, rinse very well, and then boil wine bottles—saved, bought, or scrounged from restaurants. The wine is siphoned into bottles, which are corked (with a corker and corks that have to be purchased) and stored with the bottles lying on their sides, so the liquid touches the corks to keep them moist and tight in the bottlenecks.

SWEET OR DRY

If we decide the wine is to be sweetened, we add anywhere from a quarter-pound to a pound of sugar per gallon of fluid. The sugar, as usual, is boiled for ten minutes and cooled in just enough water to dissolve. The amount of sugar is pure judgment. (Since

Corker

the first fermentation normally uses up all the sugar, the alcohol content is high enough so the yeasts are killed. Thus, most of this second sugar will stay to sweeten the wine.) A quarter-pound (a half-cup) of sugar per gallon will turn, say, clicking-dry dandelion wine to a smooth, pleasant drink to accompany fish. If we add a half-pound, it becomes a nice dessert wine to go with fresh strawberries. Three quarters of a pound turns it to a definitely sweet aperitif, a before dinner wine that's good over ice cubes.

MORE AGING   In any event, the wine should sit for another six months or a year after the second batch of sugar is added; the longer the better. There may be a slight second fermentation if the wine has not generated enough alcohol to kill off the yeast plants. This percentage varies from 12 to 14 percent, the maximum alcohol amount any wine can develop naturally. That's about twice the amount of alcohol in commercial beers and a quarter of that in a distilled whiskey.

We bottle the sweetened wines at the end of their second year's aging. In general, the longer the wine continues to age in the bottle, the better. We always save at least one bottle from every good batch to age for a full ten years. We are a way from opening one of those, but we're looking forward to the day.

By the way, if anyone tells you that home winemaking is against the law, you can tell him he's wrong. Any head of a household can make up to 250 gallons per year for home use. You can't sell it, can't distill it into brandy (though how anyone would know if you brewed up a bit of hard stuff on the stove, I'll never know), and to be strictly legal, you should obtain a form from the local Director of Internal Revenue. Frankly, except for antimoonshining activity in the South, enforcement of the archaic Prohibition era laws against home booze-making is nonexistent. After all, Sears & Roebuck is selling home winemaking kits by mail...no forms required.

Final Harvest:
Meat and Hides

11

As fall melds into early winter the brilliant New England foliage turns dull brown, then falls, and chill rains begin to alternate with days of sun that is bright, but too weak to cut through the bite in the air. Nature is girding for the bitter winter to come and though plants and cold-blooded creatures are slipping into dormancy, the warm-blooded animals are at a peak of activity. The deer, sleek from the summer's browse, are in rut, using their season of peak physical condition to guarantee the survival of the species—fawns conceived this fall will be born next spring just as the first leaves begin to bud. Squirrels are busy in the oak groves, gathering and storing the acorns that will sustain them through the months of deep snow. Partridge, pheasant, and wild turkeys rustle through the brush, gorging on seeds and wild fruits to build the layers of body fat which will protect them from the winter's cold. And deep in the woods the Canadian or varying hare is losing its summer coat of brown, the winter white growing out in spotty patches, a forecast of the snows of December.

In our barnyard too, nature's preparations for winter are under way. The chickens' appetite increases, as does their sub-skin layer of golden fat, and the rabbits are growing a denser fur. The hogs seem more voracious than ever, determined to put on their own insulating fatty layer, while the goats are growing their thick and shaggy winter coat.

Now is the time when nature has brought her creatures to top condition, flesh insulated and marbled with fat, hides, and furs, at their finest. And now is the time that this homesteader takes his share of the wild animal populations and puts away the meat and skins of all but the few farm animals that warrant being fed and housed over the winter.

Fish for Food

The first part of our fall meat harvest takes place after a few good, hard frosts, when little rims of ice form along the shore of still ponds. One day will dawn biting cold and clear to find us, spinning rods in hand, on the shore of a nearby lake or slow-moving stream. It's the time that freshwater fish go on a fall eating binge, eager to store up the fat they will need during the months of near-dormancy spent under the ice.

We aren't repeating the spring's fun fishing when we went after tackle-busting bass and landlocked salmon or tried to beguile the wily native brook trout to take a tiny dry fly. That is sport and fine sport at that. Now we are fishing for food, seeking the easy-to-catch, delicious to eat, and largely neglected panfish—bluegill, sunfish, crappie—and the delectable catfish. Panfishing is done with a small artificial ''wet'' fly tied to a casting float. Tossed into just about any patch of withered pond weed or lily pads, the fly will usually be taken instantly by an eight-to-ten-inch-long panfish. If the fly doesn't work, we resort to honest worms, dug from a compost heap.

Fall fishing seems to be best the hours just before and after dusk, so after an afternoon spent taking panfish we go after the night-feeding catfish. The tackle for ol' whiskers is a simple hand-line with a lead weight tied on one end and a hook tied in several feet up from the weight. Baited with well-aged panfish innards, these hand-lines can be relied on to produce one good catfish each quarter-hour from just about any weedy cove.

FISH-CLEANING   Back in the kitchen we make fish-cleaning an assembly-line operation. With the catfish, I use a set of flat-jawed fish skinning pliers to snip off the tail and the three barbs on top and at each side behind the gills. Next the skin is cut all around just behind the head. I put the head in the jaws of a fish-cleaning board and use the pliers to shuck the skin off. Then the head is cut off, and any innards that didn't come off with the skin are scooped out

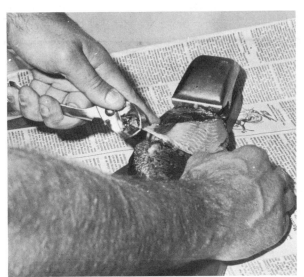

Cleaning the catch is the part of fishing most people dislike, but there's not all that much to it. With the fish clamped on the cleaning board, I pull off the skin. If it's early spring, you shouldn't forget to look for the roe sacks—small but tasty [above right]. Finally, the top half of the filet is cut free of the spine [below left]. The result is a delectible chunk of bone-free meat from a fish that a lot of people dismiss as too small to clean. And there's another like it on the fish's other side.

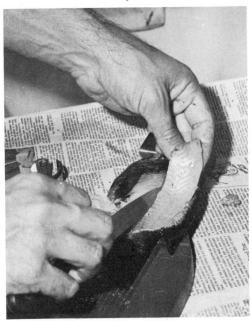

and Louise packs up the resulting "drumstick" of meat for the smoker or freezer.

Most people scale and gut small panfish and cook them bones and all. We fillet them, losing only a tiny portion of flesh in the process and saving a good deal of freezer space. I put the fish's head in the cleaning board clamp, cut the skin just behind the gills and down the back along the dorsal fin. The skin and scales peel off with one or two passes of the pliers. Then I cut away the meat with a thin filleting knife, starting at the top along the spine and slicing down over the ribs wherever the meat is thick enough to bother with. A perch will yield a cigar-shaped fillet from each side, a blue gill or other "fat" fish a sort of L-shaped fillet.

The skins and heads and all are collected in a bucket and served up for the pigs' breakfast. The meat Louise washes well, soaks in a light brine to remove any possible muddy flavor—a half hour or so for small fillets, overnight for large catfish. Then a good dinner's worth is packed into double-thick plastic bags, cold water is added—about a cup per bag—and the future fish fries are quick-frozen. The water is added to seal in flavor. Without it, many fish become dried out and tough in the freezer.

Dressing Poultry and Game Birds

Since birds are all built about the same, the process of dressing them for the table is similar whether you are dealing with a domestic Pekin or wing-shot mallard duck, a young cockerel rooster, or a ruffed grouse you spent the whole morning stalking. I'll not go into the particulars of hunting. There are whole books on gunning for upland game birds or stalking deer, and most of them are entertaining reading. And besides, as I don't like guns all that well. I get most pleasure from stalking or luring a wild animal and dealing with it on its own terms, so I prefer to use a strong **AGING GAME** slingshot or a bow and arrow to bag a wild animal.

Now, domestic poultry should be dressed and either eaten or frozen right after it's killed. Game birds, being naturally tougher, benefit from a period of naturally aging. We hang game in the feathers for from one to ten days depending on the species and temperature in the barn or cellar. This allows natural enzymes to begin the decay process just enough to tenderize the meat and allow the feathers to be pulled out easily without tearing the skin very much. This is the selfsame process that is used to make red meat edible, by the way. Without aging, a sirloin steak would taste like shoe leather. A game bird would taste worse.

Most properly aged game birds and some domestic fowl, squab and turkey in particular, can be dry plucked. For chickens, ducks, and other domestic birds that we keep around the homestead, I fire up a big old twenty-quart canner that is placed on a hearth made of flat rocks, filled three-quarters full of water from the hose, then heated up with the flame-sprayer. This gadget uses a hand pump to blow kerosene through a metal coil that's kept heated by its own flame. The kerosene vaporizes, and roars out of a small orifice in a flame that reaches 2,000° at the tip. Very handy for burning out tent caterpillars, disinfecting animal quarters, and heating up scalding pots at slaughtering time.

CHICKEN PLUCKING

When the water is hot but not boiling, I chase the birds into the henhouse, close the sliding door, and reach in for a Sunday dinner. The best way is just to grab the first leg that comes to hand. The flock will jump around, raising dust and making a big racket, but can you blame them?

There are numerous ways to kill a chicken. You can wring their necks: hold the head tight and whirl the body around till it flies off the head. The carcass will run around "like a chicken with its head off," gurgling blood and flopping for a minute or two. The same effect can be had by chopping the head off, but this takes two people unless you want to lose a lot of neck meat whacking away at a moving target. I rig up several lengths of baling twine off used hay bales, putting a fixed loop in one end and a sliding noose in the other. I put one of the chicken's legs in the noose end, draw it tight, and hang the fixed end on a nail on the grape arbor. Next I half-fill a big steel bucket with hog grain. To the handle is attached a length of clothes hanger fashioned in a hook. I punch the hook through the bird's lower beak and cut its throat just behind the lower mandible with my big stock knife. The bucket's weight keeps the bird from flopping around and the blood, which would otherwise be wasted, becomes part of the pigs' next meal.

KILLING

When the blood has stopped draining, Louise fastens a length of twine to the bird's other leg and, wearing a heavy rubber glove against a hot water burn, submerges the carcass in the hot

The bird is hung for slaughter by a cord around each leg, and a blood bucket is suspended from a hook put through the lower bill. I stretch the neck skin and simultaneously make one quick pass just below the bill with a sharp knife. The bird flaps about, but the weight of the bucket keeps it from flailing and getting blood all over. After the bird is bled out, Louise immerses it for ten seconds in not-quite-boiling water to loosen the feathers, which are then stripped off. The big wing and tail feathers come off first, followed by the body feathers, which generally peel off like a sheet. It's a good idea to check for pinfeathers before the slaughter. Otherwise you'll have to pick them out one at a time, as Louise is doing, bottom far right.

(but still not boiling) water bath. It is swished around for about six seconds and she makes sure the water gets to the base of all feathers. Then it's pulled out and held at arm's length to let water drain back into the pot, while the worst of the wet-chicken-feather-smelling steam rises out of range of her nose. Then the bird is suspended by both legs and the feather stripping commences. First to come out are the toughly-attached pinion feathers in the wings and the tail feathers. Usually the rest of the job can be done by stripping the feathers off as if they were a sheet, similar to stripping a heavy layer of suds from a plate. If the skin tears (usually on the breast), the hot water bath was too hot or long. If

feathers are hard coming out, it was too cool or too short. If the bird doesn't come clean with one pass—except for a few hairs—and has a lot of pin feathers, which are like little teeth of a plastic comb sticking out of every skin pore, the bird is moulting and the picking job will take hours. Normally by late fall all moults are over and the winter feathers are fully grown and easy to strip off. We always check a bird before killing it, however, to make sure it isn't all pin feathers under the outer cover.

The hairs that remain on the carcass can be singed off; I just make a pass with the flame-sprayer. Some breeds with dark plumage will have dark spots remaining in the hide. Unless there

With the carcasses all cleaned, I carry them into the kitchen for cutting and gutting. What with those tomatoes and zucchini in the cart, it looks like a big feed in the offing.

is a piece of feather quill that must be squeezed out, such a spot is nothing to bother about. You can cut them out for appearance's sake, or do as the commercial broiler factories do and raise only white-feathered stock. We just ignore them and they disappear when the bird is roasted to a golden brown.

Gutting and Cutting

We have a couple of tricks that make dressing birds out a snap. We'll get to them in a bit. But for a whole roasting chicken to stuff and serve for a big occasion, you have to do it the old way. It's time-consuming and even after years of practice takes us five to ten minutes a bird.

I lay the plucked and singed bird out on its back on a heavy chopping block and whack off the head and bloody half-inch end of neck with a butcher's cleaver. To remove the feet, I bend the **OFF WITH THE HEAD**

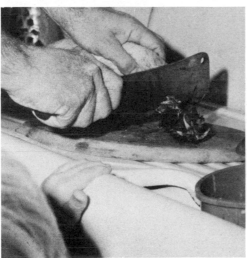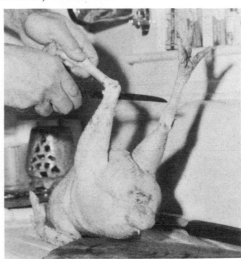

The first step in cleaning a chicken is to chop off the head and part—about a half-inch or so—of the neck. Then the legs come off. Scrubbed well and blanched till the scaly hide comes off, the legs can be boiled into a soup or gravy base that is especially prized by Oriental cooks. The small hand and the shock of hair in the lefthand photo belongs to Sam, who isn't in the least put off by what's going on.

yellow, scaly foot in each direction as far as it will go. That way you can discern the joint ends on the foot and drumstick. By cutting each side in the middle of this joint, I sever the tendons that hold the foot to the leg and it can be removed with a good twist. There is always a small bit of yellow scale left on the drumstick which can be peeled off easily.   **AND FEET**

With the bird on its back, legs pointed away, I remove the neck and craw. Then I cut the skin along the underside of the neck with a pair of tough poultry shears. The windpipe is clearly visible and when you get up to the body you'll find the craw. It's located **OUT WITH** just under the skin where the neck attaches to the body. Usually it **THE INNARDS** will be full of grain and easy to find. If empty, it is a whitish sack. The craw will pull free and with a couple of knife cuts the windpipe will come off with it. Then I cut the neck from the body, peel it free of the skin, and toss it into the giblet pot.

Next the bird is turned over and I cut out the wax gland, a little spout located on the back just forward of the tail. All the rancid-tasting yellow glandular tissue must be gotten out. Then with the bird on its knees, so to speak, I use a very sharp, pointed

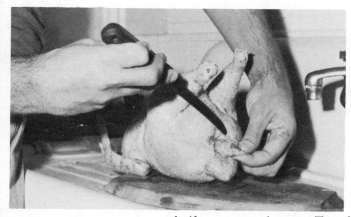

Cutting around the vent must be done carefully. A cut into the chicken's intestines will be regretted. To clean the bird conventionally, this first cut would be enlarged and the innards pulled through it.

knife to cut out the vent. To reduce the chances of cutting into the gut and getting fresh manure all over you and the bird, cut a hole about the size of a quarter with the vent in the middle, and cut in sort of a cone shape, keeping the knife blade as close to the interior walls of the body cavity—and thus away from the innards—as you can. With the shears I cut the tough yellow skin from the vent incision down to the breast bone, then open up the skin flap and use the knife to cut all around the vent incision through the thin but tough layer of whitish membrane to expose the internal **BEWARE** organs. At the same time I'm releasing an odor that isn't precisely **THE SMELL** bad, but takes some getting used to. Don't breathe deeply the first few openings. (Incidentally, as I've mentioned before, I'm not doing this to make anyone sick, rather to warn you what to expect in advance so the actual doing won't send you running for the bushes.)

There are two sharp little bones at each side of the vent. They can scratch, so I usually peel back the skin and snip them off. Then with the bird on its back, I slip my hand into the surprisingly warm body cavity, running fingers along just under the breast bone, gently separating innards from carcass. When my hand and wrist are in as far as they can go, I feel around for three very strong strings, one aiming straight up, the other two off to each side. These get snapped. Then I curl my fingers into a claw, and keeping tips pressed hard against the back and probing into any dents that come along, I very gently pull the stuff out. First time into a chicken you'll find the insides soft except for a very hard lump in the middle. That's the gizzard, and you want to get that out because the intestines are sure to come with it.

With luck the whole thing will come out in a piece and the

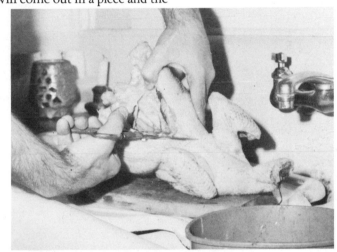

Faster and easier is cutting the bird up the back, starting at the incision around the vent, and all the way up, just along the backbone.

carcass will be clean as a whistle but for a few odd strings. Usually, though, I fail to get it all and have to go probing for the lungs. These are located in pockets up against the backbone in front and have to be scooped out. They are bright frothy pink and it's hard to mistake them. Every bit must be removed.

Finally, odd strings are pulled out, the inside washed well, the outside scrubbed even better, and the bird is ready for the table or freezer. If you have an old hen you oughtn't discard the egg-producing machinery, both the white glandular tissue and the many yolks of all sizes that will attach to it and to the back. These unhatched eggs have a marvelous taste and consistency and the kids fight over them. They are the main (visual) difference too between a real homesteader's chicken stew and what you can put together in any suburb.

The heart is easily popped free of its enclosing membrane, squeezed clear of blood, and cut from the tubes. The liver must be

FINAL CLEANUP

cut from the digestive apparatus and the small green bile sack attached to one lobe cut out. If the bile sack ruptures, discard whatever the bile gets on. The Indians used to use bile as a condiment, but for most modern people's taste it is terribly bitter. The feed-packed gizzard is cut from the other stuff and is slit where the tubes go in. It's then turned inside out, the grain is dumped out, and the tough yellow muscle lining peeled off. Once a knife pries an edge up from the muscle, the lining comes right off. The giblets, washed well, can be left with the carcass or kept separate. We freeze the livers separately, saving up for the next visit from the fresh-liver-and-onion-loving branch of Louise's family.

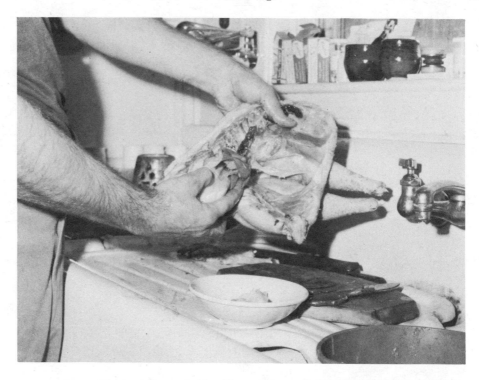

Cutting along the backbone allows the bird to be opened like a book. The insides roll out, and after removing the edible gibblets, you are done. The bird can be stuffed and roasted whole, halved, or sectioned, as you wish. Usually, we remove the legs, thighs, breasts, and wings, then cook up the rest of the carcass for stew, soup, or gravy stock.

Cleaning Poultry the Easy Way

For young chickens to be split for broiling or cut up for frying we avoid most of the hard work. I remove head and feet, then cut through the back from the neck up the spine and around the vent, and the chicken opens up and empties like an oyster. Equally simple is the sectioning method used on old hens or roosters to be stewed. The leg and thigh come free of the carcass with one cut through the skin between thigh and side, another down where the thigh joins the back, and a final twist of the hip joint. To remove leg from thigh I search out the joint much as when removing foot from leg. Find just where the joint joins and a knife slips through like butter. The only other part saved from really old birds is the breast meat, carefully skinned and filleted from the breast bone. No matter how old the bird, the white meat will be tender if sauteed till the juice runs just white in a little butter. The rest of the carcass, innards removed, goes into the stock pot with a bit of salt, carrot, and onion. Boiled slowly till the meat is drained of all flavor and the bones are falling apart, the stock is frozen to serve as the basis of many a soup or stew in the cold months to come.

CHICKEN PARTS

A separate stock pot is used for the feet. After a good scrubbing, feet are boiled for fifteen minutes. After cooling, the scales peel off and the nails come out and the bones and gelatine-like flesh is boiled up, and either added to the regular stock or kept separate to be used as the base for chicken recipes and soups from Chinese recipes. Our concentrated chicken, duck, and goose stocks will turn out meals that haven't been seen on most American dinner tables since Campbell's put your grandmother out of the soupmaking business and Colonel Sanders figured out how to substitute chunks of chemical-flavored batter for chicken.

POT STOCK

The bucket of heads, innards, cooked-out bones and all, ends the hogs' day right, with the exception of combs. These we cut off, scrub well, and add to Coq-au-Vin to convince doubters that indeed it was an ancient hen or rooster that went into this deservedly famous high point in the French cuisine. Old Chicken in Wine Sauce: here's how Louise makes it. (Try it with a store-bought chicken and you won't taste the chicken flavor at all. It requires a bird with real character born only of long experience in the loves and fights of the real chicken life.)

KEEP THE COMBS

Louise's *Coq-Au-Vin*

For each serving she first cuts up a chunk of salt pork the size of a butter patty. The pieces are simmered very slowly in butter till the fat is a warm, golden brown. Chitlins are removed and set aside.

Next, leg-and-thigh pieces of old chicken (one per light eater, two for good eaters) are dried very well, then browned in the fat, which should be heated so the skin sizzles well. Louise returns the chitlins to the pot, adds salt and pepper, covers the pot, and simmers the bird for a quarter-hour to harden up the meat so it won't lose flavor in cooking.

Then the wine goes in: three to four cups of the best red wine we can get, and usually this means a bottle of a good French burgundy or other red wine. Rich chicken stock is added to cover the meat well and she puts in two good-sized garlic cloves cut in small pieces, a quarter-teaspoon or a bit more of fresh thyme, a medium-sized tomato, seeded and chopped, and a large bay leaf. This is set to simmer in the oven for at least two hours. While so great an amount of cooking would turn a store bird to mush, an old rooster is just getting chewable in that time.

While the bird is cooking Louise sautees a skillet-full of small onions in sweet butter till they are golden brown and in a separate skillet does the same with a batch of bought mushrooms, if we don't have a good supply of wild ones pre-sauteed and stored in the freezer. (Nothing more dangerous than trying to tell people how to separate good from poison wild mushrooms in printed words, so I won't.)

When the bird is tender Louise removes it from the pot, cooks the broth down by about half, fishes out the bay leaf, spoons off floating fat, and thickens with flour mixed in a bit of water till it's moderately thin gravy. Then the meat is returned to the juice, the vegetables added, all is simmered for five minutes or so, and it's served to great applause with little new potatoes boiled in their skins, buttered and decorated with fresh parsley, plus green peas and a light salad. The dish can be let cool and sit overnight or in refrigeration for a day or two before being reheated, and if anything it improves in quality. The recipe is also fine for game birds, especially older ones.

Small Animals

"But how *can* you kill it?" is the usual question of city people when they discover that the main ingredient of the delicious rabbit stew they are enjoying was hopping around in its cage just a few hours earlier. And I'll admit that killing and dressing our first rabbit wasn't easy. There is a great deal in the homesteading experience that is less than pleasant; many of nature's ways aren't always pleasant by today's "civilized" standards. And killing warm-blooded mammals that are similar in so many ways to ourselves is the least pleasant thing we do. But isn't it somehow more honest, less hypocritical, to do your own slaughtering, rather than picking out your chops or steaks from the rows of plastic-wrapped packages in the supermarket? I think so at least.

UNPLEASANT BUT HONEST

Let's use rabbits as our major example. The killing, skinning, and dressing process is essentially the same whether the animal is a rabbit, raccoon, goat, hog, sheep, or steer. There are differences in size, of course, and in treatment of hides and cutting of the carcass. But once you've done several rabbits it is surprisingly easy to move on to larger animals.

First step is to render the animal unconscious without killing it outright. This is so it will feel no pain, but the heart will still function long enough to pump the carcass clean of blood. While animal blood is a highly nutritious food and really should be saved to thicken and flavor gravies or to make blood sausage, it ruins the flavor of meat if not completely drained out.

STUNNING

You can stun a rabbit by hitting it sharply at the base of the skull or shooting it. Easiest, most humane and sure, is to dislocate the neck. Being right-handed, I lift a rabbit by the loose skin of its back with my right hand. With the left, I grasp both hind legs at the hock, pressing the hand against my waist. The animal is draped over my right hip, my thumb against the base of the skull, fingers over the muzzle. I extend the right arm, stretching the animal tight, and then at the same time jerk hard on the legs and pull up on the muzzle as I press the base of the skull down and away with the right thumb. You can feel a distinct pop as the skull dislocates from the spine. The rabbit is instantly unconscious and incapable of movement, even if its eyes appear to be staring balefully at you.

The rabbit is hung up on a trapper's gambrel. If the blood

BLEEDING OUT

is to be saved I hang the blood bucket from its lower jaw and cut the throat just behind the lower mandible, directing the blood flow into a cheesecloth-covered pan set into the bucket. A bit of vinegar will keep the blood from congealing. If the blood is going to the hogs, I just sever the head at the spinal joint. This is not as simple as it might appear, since the loose skin makes it hard to cut in. I pull the ears down hard, stretching the skin of the back. A very sharp knife run hard just at the base of the skull will usually take the head off cleanly.

SKINNING   After the animal has bled out it is skinned. With the poultry shears I cut off the tail and the little pad of fur over the vent area. Then I slip a finger under the skin and pull it loose from the leg muscle, cutting a slit across from hock to hock. Rabbit skin is thin but the fur is maddeningly thick, tough, and surprisingly hard

to get a cutting purchase on. The skinning knife should be razor sharp and kept that way with frequent honings.

The book-educated homesteading writers tell you just to shuck the rabbit skin off after the initial cut is made, and as usual they're only partly right. The fur must be separated carefully from the leg; just pulling it will take off a lot of good leg meat. I cut the skin all around at both hocks and pull down gently, cutting the under-skin membrane where it sticks. Once the skin is down over the legs, though, it comes right off. Then I snip off the forepaws and just pull.

Next I turn the skin fur-side out again, rinse the blood from the neck, and wring it out well. Turned inside out again, the skin is put onto a stretching board, a length of six-inch-wide scrap lumber tapered to a dull point on top. The neck end goes at the

FLESHING THE SKIN

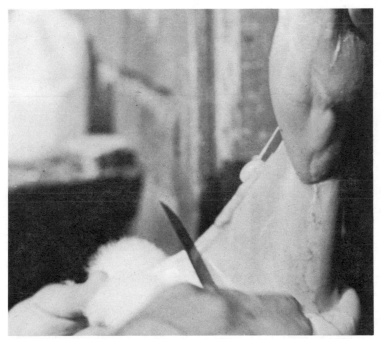

The rabbit is rendered unconscious by dislocating its neck. Next it is hung on the skinning gambrel and the throat is cut just behind the jaw to permit thorough bleeding. Then the skinning begins. The first cut, being completed at left, is made from hock to hock. The skin is then pulled free of the rear quarters. This must be done with care, using the knife where necessary, or the flesh may tear. The skinning may be begun, incidentally, before the animal is completely bled out.

pointed top; both forelegs should stick out on the same side. When it has dried a bit I "flesh" the skin—put it over a length of plumbing soil pipe, and scrape with a trapper's fleshing iron and draw knife, to remove any clinging chunks of meat or fat and the thin layer of outer membrane. This gives the hide a suede-like appearance. The skin is replaced on the board, dusted with powdered borax as a preservative, and set aside to dry. If hard frost hasn't killed all the bugs, the hides must go in a well-screened place or else they will be "blown"—infested with the eggs and larvae of the blow fly.

GUTTING   At the cleaning board I cut off the rear feet below the hock and lay the carcass on its back. A shallow cut is made in the thin abdominal membrane from the point of the breastbone to the tail. Where the knife is stopped by the pelvic arch I snip the bone joint

with the poultry shears. The majority of a rabbit's insides are filled with its digestive apparatus which will just roll out, needing only a membrane cut here or there. The terminal end of the intestine running out through the pelvic arch will usually be filled with unexpelled droppings and the stomach will be full of food, the bladder full of urine. So pull gently. Or you can do as is advisable with larger animals and withold food and water for twenty-four hours before slaughter, and have a somewhat easier cleaning job.

At the front of the lower body cavity is the liver. I cut it free of the spine and snip out the small green bile sack. Along the small of the back are the kidneys, partially buried in a waxy white leaf fat. We don't like the soapy taste of rabbit fat, what little there is of it, so I pull it all off, retaining only the kidneys. In front of the liver, dividing the chest from the abdomen, is the thin but tough

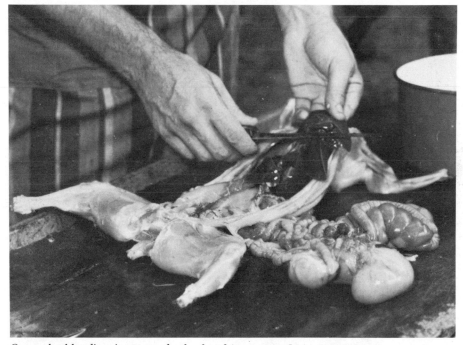

Once the bleeding is stopped, the head is removed [far left], then the forepaws [center left], and finally the skin is stripped off like a glove [near left]. Always, care is exercised to avoid tearing the flesh. With the skin off, the carcass is opened up along the abdomen and the innerds removed. In the photo above, I am careful removing the gall sack from the liver.

diaphragm. This should be cut out. Behind it are the heart and lungs which come right out along with the breathing tubes. At this point I split the breast and clean out the blood that usually collects in the neck meat.

BUTCHERING
　　　　　　Washed briefly in cold water, the carcass can be frozen as is or cut up. To cut a rabbit we first spearate the hind legs from the back. You can tell where to cut by moving the legs and watching where the muscles are attached. The main part of the back with the thick "saddle" is cut away from the upper back at the lowest ribs. Then we cut along the backbone, dividing the ribs and forelegs into two serving pieces. It's instructive to examine the carcass at this point, because each part has its precise counterpart in the larger livestock we will be considering later on. The rabbit is a miniature of just about anything that walks on four feet. Its rear

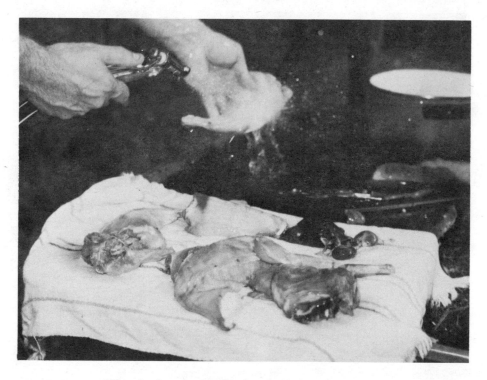

The final step in the butchering process is flushing off the sectioned rabbit and giblets.

thighs are hams in a hog, one half the saddle a crown roast of lamb. The little thin membrane that covered the stomach would be sold as breast of lamb or salted, smoked, and sliced as bacon in a hog.

Cooking Rabbit

The meat of domestic rabbit is usually described as having a flavor similar to white meat of chicken. Which is to say it doesn't have much flavor at all, and that's the truth of it. The flavor of most meats lies in the fat; meat by itself has very little flavor aside from an iron-like taste.

Some cookbooks would have you treat rabbit as chicken. We think this is disappointing. You end up with fried rabbit or rabbit stew that is not one whit different from chicken cooked the same way, since you are using the familiar chicken spices and cooking techniques. One spice that goes particularly well with rabbit is black pepper; use plenty of it in a breading and fry the meat slowly, covered for an hour, in fresh pork fat. Traditional recipes—such as haspenpfeffer or jugged hare for wild rabbit or hare—are good with both wild and domestic rabbits.

In our opinion, the best way to cook rabbit is to treat it as you would another relatively tough, tasteless meat, veal. Any of the hundreds of ways Europeans have dreamed up to serve their favorite, milk-fed veal will go grandly with rabbit. Try Rabbit Marsala or Scallopini by boning the meat and pounding it into thin steaks with a meat mallet. These thin steaks can be breaded and served as cutlets or Weiner Schnitzel. For a mild but delicious stew, make this rabbit version of blanquette de veau. I guess we should call it blanquette de lapin.

Blanquette de Lapin

Cut as much meat as you can get from a rabbit. Put the carcass in a quart of water along with four or five chopped ribs of celery with their leaves, some fresh parsley, one peeled and cut up carrot, a bay leaf, two cloves, a half-teaspoon of salt, and a big pinch each of thyme, grated nutmeg, and pepper (white pepper if you have it).

Simmer the stock base for at least two hours, then add the meat pieces and cook till tender, at least an hour more. For the last half-hour of cooking, add for each person to be served several new potatoes in their skins and as many small peeled onions as they might like.

When all is done, set meat, potatoes, and onions aside to keep warm, strain and skim the stock, and boil it down by half. Make a *roux* by mixing well three tablespoons of sweet (unsalted) butter with two tablespoons of flour and cooking it as slowly as you possibly can till it turns a medium brown and has a mouth-watering nut-like smell. Then pour in boiling stock slowly, and cook till smooth. Next remove from the heat and stir in three beaten egg yolks. They should thicken (but not curdle) without further heating. If the sauce has a raw egg flavor, cook over a very low heat, preferably a double boiler, till the eggs cook. Don't boil the stock or the eggs will scramble and ruin the consistency. Finally, check for seasoning, adding salt and pepper if needed, and beat in two to three tablespoons of fresh lemon juice *or* that amount of a very good brandy or sherry. The dish is pallid looking, so it should be accompanied with a good dark green salad, or perhaps carrots or broiled tomatoes to add color. Delicious!

Here's another recipe which may cause a few raised eyebrows. But it's consistent with the need we are all becoming more and more conscious of, to conserve nature's bounty, use everything we are provided, and waste not. It's named after the British Canadian who taught it to me, and who has a standing invitation to come down from the tundra and share the feast.

Tete de Lapin
S.W.O.T. Kendal-Ward Esq.

Don't give those rabbit heads to the hogs. Indeed, don't cut them off. Bleed the animal by making a small incision on each side of the throat, then skin out the head too. This will take some careful knife work to get out the inner ears and cut off the nose and lips. The eyes should be cut out too, which takes a strong stomach and a very sharp penknife to get at the muscles which are fixed well back into the eye sockets.

I accumulate and freeze rabbit heads, as Louise won't even look at one. But when a fellow fancier of this treat comes to visit, we send the folks with queasy stomachs off to town for a restaurant meal and we thaw the heads. All you do is roast them, covered in a bit of butter, at 350° for an hour or till the skull bones come apart easily. Served with a big bowl of black butter, thick slices of homemade bread and a salad, the brains are delicious. The tongue skins out much as a beef tongue and the true fancier will find goodies in all sorts of nooks and crannies. Beer goes well with this dish—lots of it. The first time at least.

Small Game

Much of our country offers a largely untapped supply of red meat that is uniquely available to the rural homesteader, small four-footed animals that many consider pests. I don't know how many country and suburban gardens I've seen with fences around them to fend off rabbits, groundhogs, or raccoons. Why anyone would put up a fence to keep away one of the best and most easily caught free meals you'll ever taste is beyond me.

I don't hold with shooting most small game. That's how most people get to dislike game animals. A man goes out into the cornfield with a shotgun, scares up a cottontail, and as it runs off terrified, he loads it up with lead shot, then carries it around for half a day before he bothers to cool or dress it. That evening the family dutifully gnaws its way through a meat course with the texture of a truck tire and flavor of an old boot.

To be at its tastiest an animal should be bled and its carcass cooled immediately after it is killed. You can't very well do that with a shotgun. For best eating, game should be live-trapped, then killed and treated as you would a domestic rabbit.

LIVE-TRAPPING

For most parts of the United States the day is past when we can justify using steel traps, the kind that are concealed on land to grab an animal's foot in spring-loaded jaws. Besides causing the creature unnecessary pain the traps leave an animal easy prey to inclement weather or any passing dog. Many species will gnaw or twist the caught foot off, gaining freedom but with poor chances of survival for very long. And finally, there aren't too many places left where a steel trap wouldn't be more of a threat to a passing puppy or small boy than to a raccoon or fox.

The trapping I do for pelts—a subject covered only in passing here—employs special Conibear-type traps and only those trap "sets," underwater or inside hollow logs, that are sure to get only the animal I'm after and to kill it immediately. For meat trapping in more exposed areas I use live traps, steel mesh cages with doors at each end which fall and lock when tripped by a bait pan inside the cage. The baits tend to catch anything that is around and hungry, and many are the chipmunks, wrens, and even garden snakes I've released.

Wild animals all have highly sensitive noses and are wary of human smell. To keep them free of taint our traps never come in the house or barn. When not in use they sit on top of a rabbit pen, rain or shine. A bit of rust seems only to improve their effectiveness. And when setting the traps I handle them with a pair of trapper's mittens, made from untanned rabbit skins.

Trapping Small Game

LOCATING
GAME TRAILS

First step in catching cottontails or Canadian hares is to search out their trails, rabbit runs. It is easiest to spot their tracks after an early snow, though dropping piles are relatively conspicuous in the fall woods. But once a run is found it will produce game stew for years as succeeding generations of animals use the same travel routes. The trap is baited with a chunk of cut apple, sprinkled with cider, and placed at the junction of two or three runs. I check each trap each morning, and if a particular set hasn't caught a rabbit within twenty-four hours the trap is moved and rebaited. Caught animals are dressed just as domestic rabbits, except that I wear rubber gloves to protect against possible infection from tularemia or rabbit fever. If for some reason I can't clean an animal the day it is caught I release it. Putting a wild rabbit into one of our cages could expose the domestic rabbits to tularemia or other diseases.

From time to time a rabbit set will catch a gray squirrel. Once I kept them, cleaning and skinning like a rabbit except that the tail is left on the hide. However, recent research indicates that a decline in white oaks is directly related to a decrease in the population of gray squirrels. Apparently the white oak can't propagate without the squirrels which bury acorns all over and forget where most of them are. Acorns have to be buried to grow and no acorn can bury itself. It's no surprise to learn that white oak acorns are just about the favorite food of the gray squirrel, being largely free of the bitter tannin found in acorns of other oak species. The more white oaks, the more gray squirrels, and vice versa. Since most of the white oaks were cleaned out of our woods years ago to go into whiskey barrels and clipper ships, we do our bit to bring them back by saving gray squirrels.

The gardener's worst enemy, the groundhog or woodchuck, is just about impossible to live trap in summer when it is surrounded by food. It hibernates in winter and in early spring is pretty scrawny. So the best time to get them is in the fall when they are fat from stuffing all summer, when the grasses have largely died down so they will come to a baited trap. In our fall walks through the woods we invariably stumble across the fresh dirt from several new woodchuck holes. A trap is set close to the burrow opening, baited with an ear of our late sweet corn, and the 'chuck is usually caught in a day or two.

A groundhog comes equipped with strong jaws and a set of ugly yellow front teeth which it chatters together rapidly in self defense. If it were to connect it could take a sizeable chunk out of a man's hide so it can't be handled like a rabbit. A .22 bullet between the eyes is an effective stunner if the animal is bled immediately. I prefer to open the trap and stun the animals with a marble from my hunting slingshot. This thing is a far cry from the forked stick and inner-tube elastic of my youth. It has a metal frame with a stirrup that braces against the forearm. The power comes from two lengths of surgical rubber. Using glass marbles, it is amazingly accurate up to one hundred feet. I use it mainly for archery practice; it makes me use the same muscles and coordination as the bow without having to retrieve arrows. At half-draw it is an effective but harmless deterrent to most unwanted animal visitors, ranging from dogs after a chicken to a neighbor's horses that would like to spend the whole month of August in my corn field. At full draw it will stun or kill small game at close range and a well-placed marble between the eyes of a trapped groundhog means an immediate and humane end.

I hang a stunned 'chuck by its hind feet and immediately cut off the head for quick bleeding. With a curved and round-tipped caping (skinning) knife I cut through the hide around the short tail, vent and genital area, then in a straight line down the belly to the neck. I remove the paws, then cut the hide straight across from paw to paw. I peel the hide from the carcass, using the blade to cut it free at sticking points such as along the backbone. Dressing and cutting the carcass is the same as for a rabbit except that a 'chuck has musk glands that must be removed. They are waxy, teardrop-shaped things that most wild small animals have in the small of their backs and under their legs; a ground hog has

GROUNDHOGS

STUNNING

SKINNING
AND DRESSING

them in the back and under the front legs only. (This skinning and dressing process is the same as used for beaver and raccoon except that the raccoon's tail is left on. Small fur-bearers such as mink, weasel, etc., are skinned "cased"—in one piece—the same as rabbits.)

USING THE HIDE   A woodchuck hide isn't very attractive. The sparse, ratty hair could serve no ornamental purpose imaginable. Ugly it may be, but once processed, 'chuck hide makes Grade A, No. 1 rawhide for boot laces, snowshoe webbings, and thongs for Louise's hanging planters. We don't get around to tanning till winter so the hides are stored. I pile them in a keg between layers of wood ashes and borax, mixing three parts ash to one part of the preservative. Kept wet, the mild lye from the wood ashes removes the hair while the borax retards decay and keeps the bugs away. Hides so treated can be kept almost indefinitely so long as the ash/borax mixture is kept moist. Many times I've let the 'chuck hide keg sit out through half the winter. When brought in and thawed the skins are as good as ever.

MUSKRATS   Muskrats have an unfortunate name and appearance. With their long, scaly tail they do look like a flat-faced rat though their lifestyle is more like a beaver's. A diet of succulent aquatic vegetation gives the muskrat a mild and delectable flavor and its winter coat is a dense and silky fur. But such is the stigma of an unfortunate name most of us ignore the creature. Trappers, being a terse but realistic breed, call the animal simply a 'rat; they prize the meat and do nicely selling the furs. But time was, a muskrat coat went under the name marsh hare, and today when its flesh appears in a market it is likely to be labeled marsh rabbit.

A good many of the ponds in our area were formed by blocking streams with earthen or rock and earth dams. Muskrats burrow around in the dams, weakening them, so a muskrat trapper is welcome. I tie a couple of live traps together, wire them to a pair of logs, and float the trap out onto a good 'rat pond. Baited with carrots or cabbage leaves, the traps will fill nightly till the muskrats are trapped out for that year. In a year they will be back in full force.

A muskrat is a placid animal but it can bite, so I stun them the same way as a woodchuck. After I remove the head and legs the skin comes off in a piece as does a rabbit's. I flesh the inside of the pelt, rub it well with salt, reverse and put it on a stretching board, then work a good application of powdered borax into the fur. After drying well, the hides will store for a good while before being tanned. The 'rat carcass will have musk glands under all four legs; once these are removed it can be cooked as you'd cook any game.

Cooking Game Animals

In a sturdier time my forebears and yours took small game as it came and were thankful for it. However, even properly killed and bled game will be pretty chewy and will have a sort of liver-like "gamey" flavor that is too strong for most of us moderns. The flavor is strange enough to most folks that if they tried eating fresh woodchuck, say, roasted over a bed of coals, they'd have a hard time keeping down the second bite.

To reduce the toughness, the animal should be aged in a cool place for about a week. Then to reduce gaminess it should be cut in serving pieces and soaked in a mild brine solution with some baking soda added, say, a half-cup salt to two quarts of water plus a tablespoon of soda. Game seems to benefit from freezing, up to six months. Then before cooking, marinate it for at least a couple of days in red wine or vinegar and water. A good marinade is a cup of vinegar or wine, a chopped onion, some peppercorns, a chopped garlic clove or two, a splash of olive oil, and a good dose of sage, thyme, or bay leaf, or all three. To be frank, the idea is to leach out the wild flavor and replace it with a taste that is more familiar. As time goes by you can lessen the bother of all this preparation as the family gets used to the game flavor nature intended. Or at least that's what I keep telling myself, in spite of our kids' absolute refusal to so much as taste most wild meats.

IMPROVING THE TASTE

A favorite of all ages around our place, however, is roast opossum. It can be a mite greasy, but it is very tasty, and guaranteed to have a strong effect on the city folk you serve it to. We find it's best to let on that you are serving suckling pig, at least till your more "civilized" guests have taken several good bites.

OPOSSUM

The first step is to catch a 'possum. I wire a live trap on a low tree limb and bait it with ripe meat of any kind. (Try leaving the trap on the ground and you'll have a skunk on your hands again. Once was enough for me, thanks.) A stunned 'possum's throat is cut to bleed out, though the head should be left on. The feet are removed, it's gutted completely, and I cut through the breastbone and throat to get out the windpipe and swallower. Then the carcass is soaked in two gallons of water mixed with a cup of salt and a good shovel of hardwood ashes. Left in a cool place, it should be stirred as often as you think of it till the hair slips off easily, in about two days. Then the carcass should be dehaired and scrubbed well inside and out with a tough brush. Scrape the little hairs off the tail too. The eyes are dug out, the teeth brushed, and the next step, so far as I'm concerned at least, is to pack the thing in a plastic bag and freeze it till I'm ready to face 'possum again in no less than a month's time.

ROASTING OPOSSUM

To cook, we thaw the critter and soak it in a mild marinade for a day or two. Then the skin is pricked as deep as the cooking fork can go all over and a block of wood is put in the mouth. We set the marinade to simmering on the stove top to mask the unfamiliar smell of roasting 'possum and cook it in the oven for two hours at 350°. Louise roasts the 'possum in a covered pan with a rack on the bottom. She pricks the skin continually to let out the fat, and toward the end of the cooking process bastes frequently with butter.

When the skin browns (which it will most gloriously), we put cranberries in the eyes, an apple in the mouth, give it a necklace of parsley, and make a gravy of flour, the degreased drippings, and water. Served up with boiled potatoes, a green salad, and cornbread, it is something. The legs and saddle contain a good bit of meat. So does the tail, which you eat more or less like an ear of corn. The skin is delicious if you've gotten most of the fat out; and, like a rabbit, the head can be enjoyed if you've the inclination.

Larger Animals

SUBURBAN DEER

In those parts of the country most suitable for homesteading, and certainly in the hills of New England, the only meat animal of substantial size which nature provides gratis is the whitetailed or Virginia deer. And as mentioned earlier, the species has so adapted to the changed environment of America that it is vastly more common now than when the country was first settled. In fact it has become so skilled at living in comparative safety in the proximity of man that the species has earned the nickname "suburban deer."

So wily has the deer become that it is very hard to bag by means that can be even remotely considered sporting—at least by my standards, which hold simply that a man should use the superior wits nature gave him to outsmart the deer's superior senses, speed, and knowledge of his habitat. I figure that if you can't get close enough to a deer to hit it with a thrown rock, you don't deserve venison at your table. And I'll be the first to admit that venison isn't a daily fare with us.

I'll not presume to try passing out the last word on bowhunting for a deer in a homesteading book. There are plenty of full-length books on the topic. But no amount of reading can make a successful deerslayer of anyone, even a dedicated homesteader. A bowhunter must be first and foremost a woodsman whose fascination with nature extends beyond his gardens and barnyard for more than an occasional blueberrying expedition, mushroom hunt—or deer stalk. He must know the woods intimately, know the deer and their habits: where they sleep and feed, the trails they use, and the thickets where they bed their young. He must be able to sense much of what a deer sees and hears, the movement of scent-bearing air through the trees, the minute but telling sounds and signs of nature. He must be able to distinguish the frayed twig a deer has nipped from the sharply cut nibble of a cottontail, the rustle in the leaves that means a foraging chickadee from the soft crunch of an approaching buck.

HUNTING LORE

Very little of this knowledge can be taught, and practically none can be learned from books. It hasn't been put in words; much of it can't be. It comes as the result of hours spent in the woods, much of it moving a careful step at a time, even more standing or sitting motionless till the animals have forgotten the presence of alien man. No hunger for venison or "thrill of the chase" can make you a woodsman (or since Louise knows more about the woods and its inhabitants than any hunter I've come across of late, perhaps we'd do well to use the term "woodsperson"). And the main pleasure a woods-savvy bow hunter receives comes from the hunt, the stalk, his success in outsmarting the deer on its own range and in its own terms. There is an elation felt on the release of the arrow, true. And a satisfaction in your hard-won skill if that arrow flies true and strikes home. But after the inevitable period of tracking—hours or perhaps days—when the hunter comes on the dead animal the feeling is regret, sadness at the taking of life. The let-down is tempered, however, by the knowledge that the hunt has been a fair chase, as true to nature's most elemental rules as a human can make it. So long as he follows the restraints imposed by game management legislation, as well as the self-imposed patterns of hunting behavior we call sportsmanship, which I'll admit I personally extend to an extreme by limiting the hunt to primitive weapons, the homesteader has every right to the venison on his table. So, good hunting to you.

Slaughtering and Butchering Large Animals

Killing and cutting large farm livestock and large game animals are probably the ultimate in homesteading skills, and the ones the fewest people will want to tackle. They'll have it done by a commercial custom butcher with all the necessary gear at hand. For example, in the usual American pattern of butchering a saw cut is made all down the center of the animal's backbone. A special electric saw makes this an easy job. By hand it is an hour's hard work.

We do our own slaughtering and butchering, but we avoid most of the hard work by doing it the old-time country way, similar in many respects to the way European butchers cut their meat, with much of the bone out. It is no exaggeration to state that our own meat cutters are trained to get the most bone and fat onto a chop or steak; ask a butcher if you don't believe it. We also make use of the complete carcass; heads for headcheese, stomach for haggis, and other assorted organs for sausages. Most bones are boiled for stock, then put through the shredder and added to the animal feeds. Unusable innards go to the hogs, paunch manure into a compost heap, and the hides into outer wear.

Another secret is that we use animal products in complementary ways. For example, chevon—goat meat—is very lean and what fat there is is strong tasting and sticks to your teeth. (The same is true of deer and older lambs, at least in our opinion.) So we cut most of it off, putting it in a mesh bag that hangs in the chicken pen to provide a major protein and fat component to the poultry diet. However, hogs have wonderful tasting fat, so we lard our chevon roasts with fresh pork fat. Rabbit is also defatted and is often cooked in pork fat, giving it an added fillip of flavor. The finest sausage you can think of combines sheep or hog casings (intestines), internal fat from a pig or old laying hens, and ground chevon. With a freezer you can save up one or two ingredients till the third is available.

There are differences in the ways commercial packing houses treat different animals, but when cutting the bone-free way, the differences are minimized, to the advantage of the home operator. For one thing, I skin hogs. It's a time-consuming chore but at least I have a piece of leather to show for it. In commercial operations the hog is scalded, scraped, and most of the hide and fat sold along with the meat to consumers who just throw it out. The rind on a piece of salt pork, fatback, or chunk bacon is hide and gets wasted. So too does the hide on hams, as cooked and served by most Americans. Hide on the fat parts used for rendering into soap is properly edible as chitlings, but that treat is pretty much limited to the wise countrymen of the rural South.

Preparations for Slaughter

THE DAY BEFORE

All livestock should be kept off feed and water for the twenty-four hours preceding slaughter. Their guts will be empty and the cleaning operation will be easier; a full paunch will roll out on you and make a real mess. I try to withold from all stock large and small except the goats. It is hard enough to slaughter a goat as it is, but isolating it from the herd, without feed and water to boot, sets up such a pitiful bleating that the fateful day never seems to arrive. Pigs are just shoed into a corner of the lot and kept in with a special length of electric fence. Lambs I bring down to the all-purpose pen where I clip the fleece from the lower part of the neck to expose the hide, making the sticking job easier the following day.

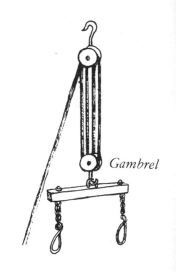

Gambrel

Next morning, if the weather reports promise at least two more days of weather in the thirty-five to forty degree range, I assemble the slaughtering apparatus. It's surprising how good an excuse the weather can be to put off slaughtering; our first pig came close to dying of old age since all fall I read the weather reports as either forecasting too warm weather (over forty degrees) or too cold (below freezing). First, you need something to stun the beast, that is, unless you are up to listening to the squeals or bleats of a stuck animal. I'm not. A well-placed .22 caliber bullet will do, as will the concussion of a shotgun shell with the shot or slug removed. Small goats, pigs, and lambs can be lifted to cleaning

EQUIPMENT

Caping knife

hooks; for mature hogs I use the block-and-tackle fence stretcher to raise the carcass on a gambrel—a length of wood with an eye bolt in the middle of one side and chains at either end, to which the hind legs are attached with wire nooses.

For skinning I use a knife with a caping blade; it curves up at the tip and is rounded so it will cut skins free without piercing through. You need a carpenter's crosscut saw or a hacksaw with a meat blade to cut bone, a heavy cleaver, and an assortment of thin-bladed boning knives. Two buckets are necessary, one for blood and the other for the innards. And finally you must have a table long enough to hold the carcass, some clean cloths, and water. If dressing hogs the conventional way you'll also need a supply of hot water, either a fifty-five-gallon oil drum half-full of scalding water to dip the carcass into, or a tub of hot water and some sort of cloths—burlap feed sacks for example—to dip into the water and lay on the hog to loosen the bristles.

STUNNING

To stun hogs and lambs I press the muzzle of a .22 pistol hard against the skull between and about an inch above the eyes, aiming shallow so the bullet will travel through the brain, not down and through the muzzle. Goats have a very thick front skull bone, the better to butt with, and so must be shot in the back of the head, centerline of the skull with the muzzle placed between the ears and the barrel aimed so the bullet will travel in a line that would emerge about midway back along the lower jaw. Put another way, the brain of these animals is about the size of your fist and lies in the rear dome of the skull, most of which is devoted

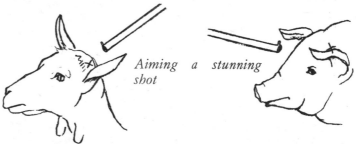

Aiming a stunning shot

to eating and smelling equipment. The line between the eye sockets defines the lower and front face of the brain, the ear canals the rear of the brain, about midway down. A bullet placed in mid-brain will render the animal instantly unconscious.

BLEEDING OUT

Such a wound will be fatal in a short time, so it is necessary to open a vein immediately to let the heart pump out the blood. A ''fiery'' carcass with blood still in it is practically inedible if it doesn't spoil, which it will if not super-cooled right away.

Sticking a pig or hog is really unpleasant; you have to cut into the chest cavity. I hoist the animal up so I can get a cheesecloth-covered bucket under the head. Then with the caping blade I make a cut in the neck skin more or less from a point midway between the front legs to the rear of the lower jaw. I make a series of shallow cuts through the neck and chest fat, which can be several inches thick in an old sow. When I hit bone between the forelegs, I cut forward till the bone runs out. This is the front edge of the hog's breastbone and the blood vessels that should be cut lie directly under the knife blade. The best way to get to them is to slip the blade up under the breastbone and into the chest cavity, then make sawing motions down and back toward the hog's head. If the cuts are made well the blood should spurt out around your wrist. (I said this was pretty unpleasant, didn't I?) If it doesn't work, you've missed the artery, and ought to repeat the cut, angling the knife slightly to one side, then the other if necessary.

Bleeding sheep and goats is a lot easier; I cut into the throat just behind the jawbones, getting clear into the neckbone. You must be sure not to cut the air passage though, or the animal may suffocate before it bleeds out.

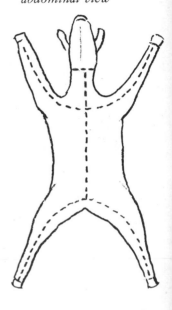

Basic skinning cuts; abdominal view

Cleaning and Skinning

The carcasses must be cooled immediately once the blood has stopped flowing. First step with sheep is to skin them out; with other animals cleaning comes first. In all cases the procedure is pretty much the same though each hide acts differently and the internal structure looks different.

In cleaning I start at the rear, cutting around the anus in a sort of inverted funnel shape. Once the end of the intestine is free, I pull it out an inch or so and tie a piece of cord securely around it. Then I make a cut through the hide from lung to throat, around both sides of udders or male parts. In hogs there is a lot of belly fat to cut through also. Next I cut into the belly cavity at the rear of the rib cage and using the round-tipped caping blade and not going in deep so as to avoid cutting into intestines, I cut on up till the knife hits pelvic bone.

Once the belly is split the paunch and intestines will fall out. If full of feed and water it will split or tear loose. When

CLEANING

cleaning a well-fed goat, I have a small table pulled up to the carcass for the paunch to rest on. Cuts are made through skin and belly muscle around udders or penis and testicles. Then, removing the insides is a matter of pulling gently, cutting strings where a piece won't tear loose. The end of the intestine should be worked down through the pelvis. If it sticks, the pelvic arch can be split with a few judicious cracks with the butcher knife and the way eased.

The bladder may be partly full; it gets handled with care, the ureter tied off if it is bursting full. Kidneys peel right off along with the fat pads they lie in. I usually cut the liver free of the gut as

My helper Kip Corey with the basic tools: a boning knife, a curved caping knife for skinning, a heavy butcher knife for splitting bone, a pen knife, a pistol for stunning, a length of rope for hoisting and a bit of twine for tying off the bung. At right, the goat is distracted with some grain, and shot in the back of the skull. Once stunned, the animal is immediately hoisted up and cut in the throat just behind the jaw to bleed out.

I get to it, cutting out the gall bladder at the same time. All the other parts are rolled into the gut tub and before being cut free the tube that goes from stomach into the chest cavity is tied off at the gut if the animal contains any food, to keep the paunch contents from getting out.

Then I use the large butcher knife to split the breastbone; on older hogs and aged goats the bone is well developed and I need a small hammer to tap the knife through. The whitish sheet of the diaphragm is split and the heart and lungs plus the large blood vessels are cut and pulled free and added to the offal bucket. Finally I go over the body interior, removing the layers of internal fat;

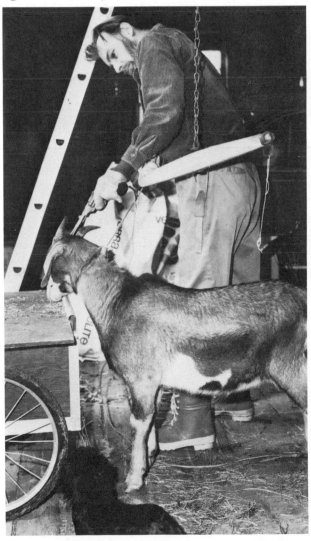

there is a particularly large amount in a hog. Leaf fat, it's called, and if left in will retard cooling, perhaps enough to cause spoilage.

SKINNING   I always skin animals right after they are cleaned; some advocate leaving them, deer in particular, in the hide to age, but it seems to me that without a hide the meat will cool down a great deal faster. With a hog skinning is pretty much all knife work. I make slits from hock to hock of each pair of legs, around the hide at the hock, then start cutting away with the caping blade. It's a good hour's work to get a relatively fat-free hide and I take it off right down to the ears. The raw inner surface is salted well with coarse rock salt, rolled and kept cool in the cellar till I get around to tanning.

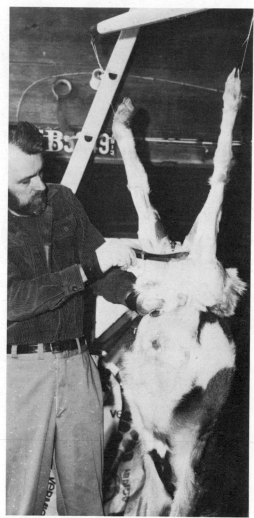

We begin skinning immediately. The first cut is made hock to hock and the hide is stripped from the hind quarters. At left I am pulling the hide with one hand and cutting shallowly where necessary with the caping blade. Kip and I continue that procedure until the hide is completely removed. Then, after the anus is cut free and tied off, I use the butcher knife and a hammer split the pelvis. With a gloved hand protecting the innards, I then slit open the abdomen.

Goats and deer are easier to skin. The same cuts are made, but once pulled free of the legs, the hides can be pulled off with just a few cuts along the backbone and around the rump. These hides are either salted or put under rocks in the springhouse well in the cellar floor. In a week or two the soaking and very slight decay process loosens the hair and outer skin covering, making the dehairing process a snap.

Skinning sheep (as said above, done before the cleaning) is a similar process, the only difference being the difficulty of cutting through fleece and skin without cutting into the meat. I pull the skin up and cut a chunk off on the inner side of each leg at the hock. Then I push my finger in, freeing the hide from the meat for

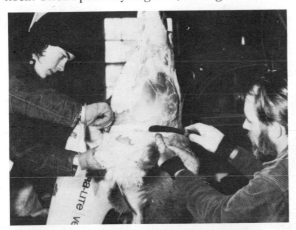

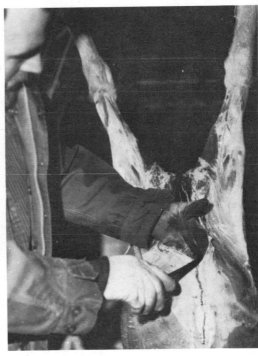

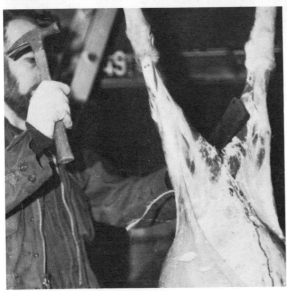

an inch or so, cut that piece, and proceed in the same hock to hock cuts, a cut down the length of the belly and another around the head behind the ears.

In all skinning you want to leave as much flesh and fat on the carcass as you can, and keep from cutting through the hide or into the flesh. Removing a pigskin, you are in effect shaving the hide off the fat. I cut a small flap free at one of the rear hocks, then keeping the hide taut, run the caping blade in an arc motion, cutting off perhaps a quarter- or half-inch section with each pass. It is a slow business but the hide comes off clean and the later tanning process goes faster. With other animals the main danger is

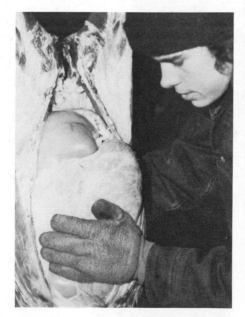

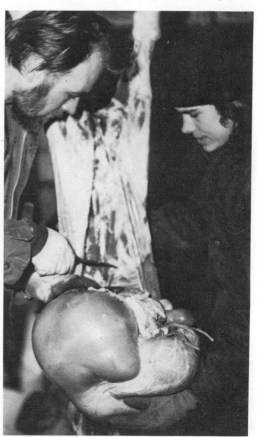

With the abdomen cut open, the entrails will begin to fall out. Kip lowers them gently into the waiting garden cart, while I cut wherever necessary to completely free them of the carcass. We save some of the organs—like the kidneys and liver—while the rest goes to the hogs.

that you can tear through the fell, the membrane between hide and flesh, and pull of a strip of meat along with the skin. To avoid this I loosen a corner, then holding the free hide in one hand I go along using the fingers of the other hand to pull the flesh from the hide. The places where the hide sticks I use the knife, cutting shallowly and carefully. This too is a slow job but the results are worth it.

Finally I prop the body cavity open with sticks and hang the carcasses to cool. If temperatures are around thirty-five degrees they just get hauled up into the rafters of the barn away from the animals. If we've not had a fly-killing frost or if the weather is either too warm or cold, I pack the carcasses into the old

COOLING
THE CARCASS

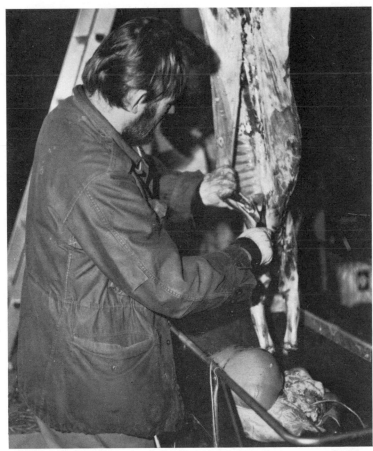

Using the butcher knife, the breastbone, then the neck, is split and the windpipe and blood vessels are removed. Finally, the chest and abdominal cavity is propped open with a stick to hasten cooling.

refrigerator that held (and may still contain) our earlier lettuce crops. Hogs are butchered the next day, other animals aged for a week or more.

SORTING THE INNARDS   Heart, liver, and kidneys are cleaned, sliced, and unless they are going into headcheese or sausage, are frozen immediately. The stomach (or stomachs in ruminants) are cut free, the contents squeezed out into a compost pile, and they are cleaned with the garden hose. Same goes for the intestine nearest the stomach; it is cut into three-foot lengths and cleaned. Once in the house the stomach and intestine pieces are turned inside out, scraped and scrubbed well and left to soak in brine (five pounds of salt to two

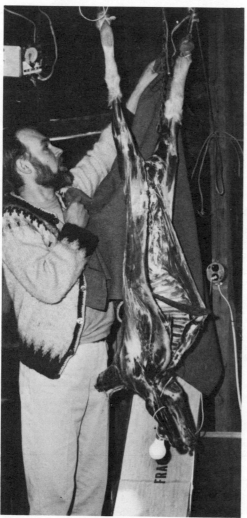

The carcass is aged about two weeks in the unheated barn, provided temperatures hold to about forty degrees for that time. If an overnight freeze threatens, I wrap the carcass with an old blanket and a light bulb to keep it from freezing. After the carcass is sufficiently aged, we butcher it. At right, Kip and I have already removed the loin and backstrap and have cut out the interrib strips for sausage. We are removing the shoulders and fores from each side, and will next turn to the neck and hams.

gallons of water) for a week. Then the pieces are removed, cleaned, turned rightside out and soaked in a fresh batch of brine for two more weeks.

Other useful organs include the pancreas, a long, grayish gland located on the stomach. Cleaned of its covering fiber and fried in butter it becomes the delicacy, sweetbreads. The bladder can be emptied and cleaned, one tube tied and the other used to inflate it. Tie the other tube off, fasten the bladder to a stick, and after it has dried you have an unusual child's toy. Put some beans in it for a good rattle; this was one of the tools of the trade of the court jesters of medieval Europe, believe it or not. Or washed very

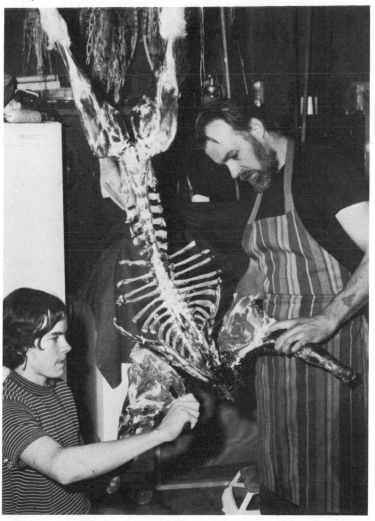

well, a bladder can be used to store wine, vinegar, or oil.

The bright pink lungs (lights) are inedible but I always cut up one into very small chunks and freeze them in ice cube trays. Thawed and tossed into a trout stream early in the year the lights show up well and spring trout love them. You can sometimes lure the trout into rising to the lights, then when they have betrayed their general location, you can try throwing some artificial lures their way.

Once all this has been put to use there isn't a great deal of offal left. What there is goes to the pigs.

Butchering the Homestead Way

BUTCHERING BY DISSECTION

Large game and farm animals are all constructed pretty much the same way. A T-bone steak in a beef steer is the same muscle as a rib lamp chop or a pork chop or the backstrap of a deer. What we do, really, is dissect the animals, separating out the major muscles rather than sawing pieces up with bone in. First I remove the head by cutting all around the neck at the skull base. The head is held on mainly by muscles and once they are severed, it can be removed by twisting a few times. I cut through the bottom and both sides of the mouth and cut out the tongue, which you'll find extends well back into the throat. All but hogs' heads are discarded as there isn't much to them but the brain which can be removed only with a lot of chiseling of the skull. It isn't very big, none of us much likes to eat brains, and the one time I tried to use them in making buckskin I found the bother more work than it was worth. So most heads go deep into a compost pile.

A Few Words on Hogs

Cutting up a hog the homestead way takes experience, preferably in butchering another kind of animal that doesn't hide its muscles under several inches of fat. However, the procedure is the same as illustrated above in cutting a young goat. If the first animal you will be butchering is to be a hog, and if you have no experience or experienced help and/or you want to cut a hog the conventional way, I suggest you send twenty cents to the Superintendent of Documents, U.S. Government Printing Office,

Washington, D.C. and ask for Farmer's Bulletin No. 2138, *Slaughtering, Cutting, and Processing Pork on the Farm*. It gives information on curing and smoking bacon and hams, pickling feet and turning a hog into familiar looking cuts.

There is nothing very difficult about salt and sugar curing and smoking pork. But this is an outdated way of preserving meat, to be honest about it, and we find freezing easier and prefer fresh pork to cured for several reasons. For one thing, home cured hams aren't the pink, juicy, tender things you find in a supermarket, that is, if you use chemical-free preservatives. Commercial hams and bacon get their appetizing redness from sodium nitrate and sodium nitrite which even the U.S.D.A. has finally admitted can cause health problems. Without the chemicals the product is a dead gray and unappealing. And to be really sure the meat is salted enough so that it won't spoil, you should use a pound of salt for every ten pounds of trimmed meat. The end product will keep but will be unpalatable unless it is cooked in several changes of water, which is a bore, removes a lot of nutrients, and most of the meat taste.

High salt content and gray color aren't so much a problem in bacon, which is cured and smoked ''sow belly,'' in reality the thin muscle sheath and thick fat layer that comprise the entirety of a hog's flanks and chest as well as the belly. We do make our own bacon; the ''belly'' starts where the loin meat gets too thin to amount to much. The two sides are trimmed square or rectangular, and the rectangles cut into five equal blocks about two pounds each. The bacon goes into a big canning kettle and over it I pour two gallons of water in which are dissolved four pounds of flaked pickling salt and a pound of dark brown sugar. The pickle goes on cold, the meat is held down with a weighted rock, and the kettle goes into the cold cellar where the temperature is around forty degrees. After a week the meat is removed, checked for spoilage (which is cut off if found), and put back in with a fresh batch of pickle. After another week or week and a half the pickling is done.

Smoking

Most of the homesteading books show a copy of a ''one-hog'' type smoker that is illustration BN-6156 in the U.S.D.A. booklet. It uses a fifty-gallon wooden barrel connected to a fire pit by a ten- to twelve-foot-long trench. Where anyone

PRESERVING PORK

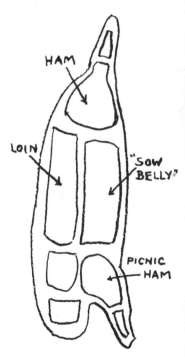

Basic pork cuts

Our meat smoking setup

could get a fifty-gallon wooden barrel these days other than for a lot of dollars at an antique store is beyond me. An old refrigerator or automatic washer or dryer cabinet garnered from the dump is more like it. But we don't even bother with that. For a few slabs of bacon I put a wooden box over the chimney of the backyard fireplace, string the bacon on stout cord that is run through holes in the box sides, and keep a slow hardwood fire going for several days. Hickory is the best-known smoking wood but any good hardwood will do. Usually we save up trimmings from the fruit trees, augmented when necessary by cuttings from one or another wild apple tree. The fireplace is cleaned well, and the meat is given an all-fruitwood smoke. The flavor is delightful and the resulting ash serves a dual purpose. I collect it, sieve it through a fine mesh to remove unburned pieces, and put it into an old kettle with the bottom rusted out so it leaks in several places. This goes over another pot, not quite so rusty, and I pour water in, adding more till the ashes are soaked and water flows through. We add a quart of water now and then for a month or so, stirring the ashes up. If the weather is still mild the pots sit out and the fall rains add to the water supply. In a month or two the lye will have leached out of the ashes. We let most of the water in the bottom bucket evaporate, then bottle what's left till we feel like showing off to some visitors and making soap from rendered hog fat. The leached ash is dried and pulverized fine. Louise uses it as a natural glaze for her stoneware pottery.

Hogs Heads, Feet and Tails

Finally, there is nothing left a hog's head, some twenty pounds of it, the feet, cut off where the joint is, and the tail. There is a lot of good meat there, but in pretty unappetizing form. I promised the British friend who shared front ends of rabbits with me that someday I'll dehair a pig's head and we will have a roast boar's head for Christmas. That's a promise not yet kept and if Louise has her way it never will be. But here's what we have done.

GETTING THE HEAD APART   First, the parts are skinned and the "trotters" or hooves of the feet chopped free with a hand ax. I skin out the head too, cutting off the ears which are nothing but bristle and coring out the ear canal which contains hairs, among other unpleasant things. The end of the snout is sliced off and the "rooter" sliced open to remove the nasal passages. Teeth are pulled, the mouth is scrubbed out well, and the eyeballs are removed. To do this I push a knife down on one side. The eye will swivel around, revealing the white muscles on the far side. These are cut with a very sharp knife. Once a few muscles are cut you can get a finger in and manipulate the eyeball. (I understand that for many years the Bedouin Arabs of the African deserts insisted that visiting foreigners join them in a feast, then offered them the eyeballs of the main meat course, claiming that they were a prime treat. It was a most effective way to keep foreigners away—worked for centuries. Of course, a Bedouin Arab wouldn't be caught dead eating eyeballs, or so I understand.)

The head and other parts are put in a large kettle, covered with water, salted lightly, and simmered slowly for a day or so, or till the bones have become unstuck—the jawbone split and the skull plates easy to pry apart. Then we let it chill, scoop most of the fat off the top, and go through the jellied soup picking out the more attractive bits of meat, discarding bones and hunks of gristle. The brain is quartered, the outer membrane removed, and meat is added to the other good parts. Finally the soup is reheated and strained into the meat pot, this to get out any errant hairs.

HEADCHEESE To make a lunch meat sort of thing, souse or headcheese, we grind up the meat parts along with the heart and kidneys that were cooked separately and the water discarded. The ground meat is added to the broth, and we put in a teaspoon or more of sage or two fresh sage leaves, several good dollops of fresh-ground black pepper, a couple of chopped small onions, a crushed clove or two of garlic, and either a teaspoon of vinegar or a quarter-cup of cut-up home-pickled hot peppers. This is cooked slowly till the broth is reduced by half at least. Turned into bread pans to harden, chilled and sliced, it tastes better than anything Armour Star calls luncheon meat.

SCRAPPLE To make scrapple, the Pennsylvania Dutch breakfast treat, we do as above, omitting the organ meats, the garlic and peppers or vinegar. When the final cooking begins we add stone-ground corn meal. You can put in a bit, cook till it has thickened, then add more and more in small amounts till the bubbles come up like miniature volcanos and it is threatening to stick. Or measure the liquid and add one fifth the amount of corn meal and stir till it's done. Hardened, then cut in slices, fried in fresh butter and served with maple syrup, hot-from-the-nest eggs and fresh bread, it is some kind of breakfast.

SAUSAGE

Once they have been through the pickling process, the stomach and intestines are scraped and washed well inside and out, then frozen or made into sausages. First we grind meat chunks and fat in two to one proportions, using chicken or pork fat with the lean meat animals. Usually this first grinding is done with the coarse plate of the electric grinder. The meat is mixed well with salt, about a teaspoon per pound of meat, plus spices. For pure pork sausage we add nothing but salt and it's great. Sausage from the stronger tasting meats needs spices. We like a lot of garlic, black pepper, and a bit of rosemary with lamb. Chevon the same, but with sage and red pepper added in. For deer sausage the same too, except the pepper goes in as whole corns and we add dried juniper berries and/or black currants, and to make a sort of mincemeat, sugar and lots of dried apples crushed into chips. Except for the salt, I can't give any proportions since each batch is done as seems best at the time and each is unique.

Once the spices are added, we put the mixture into the bowl of the big electric mixer and stir it with the dough hook. For a smoother end product we sometimes add in water, up to a cup and a half or two cups for each five pounds of meat. Finally, the mix is stuffed into stomachs or casings with the sausage stuffer attachment to the mixture. This is a funnel that fits on where the meat comes out. You run a length of casing onto the funnel, tie the open end, then run the sausage through the grinder with the fine grinding plate attached, packing it tight into the casing. When full the other end is tied and it can be divided into whatever lengths are desired by tying a piece of string on tight every so often along the sausage.

STRETCH IT WITH CEREALS

Sausage can be stretched by adding cereals. Haggis is a Scottish favorite of oatmeal, lamb, or mutton stuffed into a sheep's stomach. With pork fat, salt, pepper, and a good spice selection added it can be a real treat with chevon too. We add in rehydrated dried apple slices and roast a haggis in a moderate oven till the outer stomach is golden brown. It takes a lot of basting as the fat renders out. Roasted whole grains of almost any kind go well in sausages.

If you increase the salt content by about a quarter and let the meat cure in a cool place overnight before stuffing, sausage can be smoked and kept for several months so long as storage temperatures stay around thirty-five degrees. Smoking is done just as we smoke bacon. Sausages can also be frozen and kept for months. And finally, a highly spiced sausage, deer or chevon with plenty of grains and berries added, can be made into a pemmican that will last just about forever. The sausages, made from the smallest of the small intestines, are steamed for a half hour till cooked through, but not so that much fat is rendered out. Then they are dried, hanging from the racks of the oven which is set at its lowest temperature, 140°, and the door kept cracked so heat and moisture can escape. They will dry to a leathery texture in a couple of days, nearly bone-dry in double the time. Either way they are a hunter's delight, practically weightless and a meal in themselves that can be chewed on the go.

12

Bedding Down for the Winter

By the time the large livestock are in the freezers, late fall has given way to winter's cold. Any strong wind sucks away the warmth given off by the little wood stove in the kitchen, which means it's time to button up and prepare for the long winter. First comes us. Storm sash and doors go on and old hay bales are piled against the house to keep cold air from sneaking in between the cut granite foundation and the sills. I take the kitchen stovepipe apart and brush out the accumulated soot; the pipe radiates almost as much heat as the stove and too much soot buildup could produce a dangerous flue fire.

THE BEES   Then it's up to the beehives with more spoiled hay. The bales are piled around and on top, leaving only the dime-sized top entrance hole exposed. Any newly planted fruit trees receive their winter mouse-proofing, a big tin can with both ends out that is split, then coiled around the base of the trunk with about a third of its length pushed into the soil. The can will keep the hungry meadow mice that burrow around under the snow in winter from girdling the tree. Also, on the trees near the two sections of hill that are a favorite tobogganing area, I wrap several lengths of bright red yarn. It will show vividly against the coming snow and warn the sledders to steer clear.

In the henhouse a length of feed sack is stapled over the window and I plug in the water warmer. Feed sacks also go over the inside open part of the goats' stall and I close and latch the swinging doors to their screened window. After I check to make sure all barn windows are closed tight, the gardening equipment receives its end-of-season going over. Hoses are emptied, coiled, and hung on a spike on the back barn wall. All hand tools are gathered in, soiled knocked off, and wood handles given a rubbing with linseed oil, the metal parts a light coat of engine oil. Then they are hung on the wall of the ell. The wheelbarrow and garden carts are pulled in and stored away in odd spots, then I winterize the powered equipment.

Winterizing the tillers.

All engines are started, carburetors are adjusted, and I let the fuel tanks run dry. Then I drain the crankcase oil, remove the sparkplugs, and squirt a small amount of light engine oil into the cylinders. A few pulls on starting cords will coat cylinders and pistons with enough oil to prevent rusting. Everything is pushed into storage but the big rear-tined tiller. It comes equipped with a plow blade that is all the snow removal equipment we need, solong as I don't let the white stuff get more than a foot high before plowing. I put the blade on, then fill the engine crankcase with light No. 10 weight oil, not heavy enough to protect the engine during prolonged summer tilling sessions but good for easy winter starts and a half-hour or so of snow plowing after every nor'easter. I fill the tank with fresh gas, turn the fuel mixture adjustment to be just a bit richer—we find this helps with winter starts—and pull the machine into a corner at the front of the barn to await the first major snowstorm.

GETTING IN FIREWOOD

The next step is to get in the three or four cords of firewood that will provide a major part of our heating and cooking fuel for the winter. Time was we could buy an eight-by-four-by-four-foot cord for twenty dollars from a firewood dealer. With the shortage and skyrocketing prices of petroleum fuels, however, wood now sells for four or five times as much, making it a highly "profitable" use of time for us to gather our own. First the chain saw is tuned up and the chain sharpened, the teeth with a little rat-tail file, the spacers between the teeth pared down a bit with a flat file. The saw, files, and cans of the gasoline/oil fuel mixture and engine oil for the chain oiler, plus the other gear, are loaded into the truck which I put into four-wheel drive and we start off into the woods.

So far we have supplied our needs with naturally dead wood and a few unwanted live trees which I girdle in very early spring and let dry and cure on the stump, ready for felling and splitting the next fall. The Dutch elm disease is doing its tragic work in our area, and though elm is fairly light, thus hard to split and not a very good heat producer, it is plentiful. I try to spot infected trees before they die off completely, and dry out and lose much of their heat content. You can tell a tree is doomed if a large proportion of its branches fail to leaf out in spring; these I girdle with the ax and by late fall they will be fairly well cured. I'd never kill a healthy elm. A small percentage of the trees seem to be naturally immune to the disease and if these are permitted to stand and propagate, we may once again be able to enjoy this loveliest of American shade trees.

BEST WOODS FOR HEAT   Good firewood comes from maple, any native nut or fruit tree, and oak, among the more common species in our locale. Beech is good, but we don't cut them out unless they are dead. The nuts are a major food source for wildlife, probably more important than hickory, our most common nut tree, since a mature beech produces tremendous quantities of nuts which are easy for a squirrel to shell, and they come on every year. The hickory and butternut crops vary greatly with the weather. Both white and yellow birch are fair-to-middling firewoods with the same disadvantages as elm. Poplar, aspen, sumac, and the other very soft wooded trees are not worth cutting down. Neither are the evergreens, except one small white pine a year for kindling. Not that they don't burn well; if anything, they burn too well. The problems of using them in stoves are two. First, a pitchy wood will go up so fast and generate so much heat in so short a time the stove can overheat. And in any fire the evergreens' aromatic oils fail to burn completely so that a layer of tar is deposited in the stovepipe or flue. In time it is bound to catch fire, particularly if you let the stove get going too briskly. A flue fire is a terrifying thing—sounds as though you are inside a hurricane, they tell me. We have never had one, but just in case, there is a big-capacity chemical fire extinguisher near each stove. If the flue ever caught I would douse the stove fire, then remove a cooking lid and let the extinguishing fog go directly into the chimney.

Getting in the Wood

Probably our favorite firewood comes from the seedling apple trees that I have been cleaning out of the woods, mainly to remove breeding places for fruit tree pests. Next comes hickory. It's seldom you see a naturally dead hickory, but probably because they sprouted from squirrel-planted nuts, trees are often found growing in closely packed groups. Natural competition for sunlight would eventually kill all but one of the trees, so I have no qualms about taking all but the srongest from each group I find. Another favorite is maple, certainly the commonest hardwood in our area. Most maple firewood comes from the huge centuries-old trees that line the roads of so many country roads in New England. The main trunks are usually hollow, providing nests for many kinds of wild creatures, so we'd never take out a complete tree. Besides, with trunks up to five feet through I don't know how I'd handle one.

However, these trees are all in an advanced stage of old age and are literally falling apart. After each high wind or heavy ice storm I'll find one or two huge main branches that have been split off the main trunk. After a summer of sun drying they are ready to be cut and split into twelve-inch-long stove wood.

Over the years we've acquired a fairly complete small-scale logging outfit. The most recent addition is the electric

EQUIPMENT

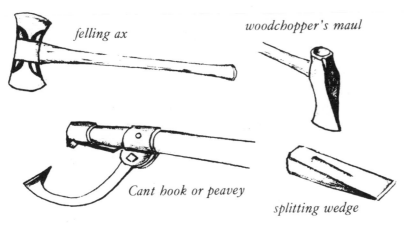

felling ax

woodchopper's maul

Cant hook or peavey

splitting wedge

winch on the rear of the truck. It will snake logs up to the truck if I can't get the truck to the wood, and will get the truck unstuck if I get over-ambitious. The tool used the most is the gas-powered chain saw. We've had several, have had bad experiences more than once, and I have a few suggestions to anyone buying a saw.

First, get a brand name you recognize; McCulloch, Homelite, Remington, Poulans, and several others are all good quality. Get the most powerful one you can afford; ours is a medium-sized Remington bought some years ago for occasional use. In long use it is just too small; it overheats and loses power.

Sharpening the ax

Too dull

Almost

Just right

Too sharp

No fault of the saw. In time, and particularly if we are able to convert the central heating to wood, I want to get a professional logger's saw that can go all day. I'd advise against getting one of these little lightweight gas or electric chain saws that are advertised heavily. Fine for occasional removal of a small limb or two, but worthless for cutting a meaningful supply of firewood. And finally, be sure to get the best quality chain offered and a cutting bar (the flat thing the chain goes around) with a roller tip. The movable tip needs greasing before each use but it reduces friction, so cuts down on wear and makes cutting easier.

Using a chain saw is easy. The first rule is to hold on tight. If it slips you can lose a leg. Never let the chain touch the ground. Just one nip of earth or a rock and you'll have to completely resharpen the cutting teeth. As it is, it's a good idea to pass the files over the chain frequently while cutting just to keep it at peak sharpness. You should never have to force a chain saw. Just hold on and let the teeth do the cutting. If you have to push down, if the saw is making smoke, or if you see a brown sap build-up on the teeth back, it's time to sharpen the chain.

In cutting a standing tree the first thing I do is go over all the branches I can reach, first cutting off all branchlets smaller than an inch through. Then, starting at the tip, I cut the branches into foot-lengths. Branches from an inch to half-inch thickness are cut up into foot-lengths with the heavy-duty rachet limb looper.

PREPARING A TREE FOR FELLING

Thinner pieces are usually let lie, though if I've received permission to cut wood on someone else's land I will gather the largest, snip them up into twelve-inch lengths, and bind them into six-inch-thick faggots with old baling twine. The heat such a bundle generates is hardly worth the effort of gathering if solid wood is available, but they make fine kindling and I don't like to leave slash around on other folks' land, even if it is only the little stuff that will decay in a year or two.

To fell a standing tree I first hold a pencil up, grasping gently with finger and thumb so it will hang vertically. Comparing it against the tree trunk I can tell for sure how the tree leans. It's necessary to evaluate limb placement too, as a heavy limb can take the tree with it no matter how the trunk leans. Once I've decided which direction the tree will go, I check to be sure its fall won't harm another, more valuable tree, telephone line, or whatever. If such damage seems probable and the tree is easily climbable I'll go up and cut out the top or whatever limbs might cause the damage. To remove limbs I make about an inch deep cut in the bottom of the branch next to the trunk, then cut through from the top. This removes the limb cleanly; cut from the top only and the wood will fall before you cut through, leaving a "hinge" at the bottom that will need more cutting or perhaps will rip off a great strip of bark as it goes.

To remove a limb, first cut from the bottom,

then from the top.

If you cut only from the top, you'll get a hinge on the bottom.

Topping a tall tree is dangerous and I wouldn't attempt it. I'll take on one of the old apple trees though, or an adolescent maple if the trunk branches in two not too far from the ground. To top out a main branch or trunk that is leaning at a good angle, the procedure is the same as cutting off a smaller limb. If the amount of lean is less, say, than forty-five degrees, I follow the procedure for felling, as described next.

TOPPING

To top a tree, you notch it on the side facing the direction of fall,

then cut down into the notch from the opposite side.

Felling and Bucking Large Trees

Once I've judged the direction of lean for a large tree, the next step is to cut out a notch low in the trunk on the side facing the direction of fall. If the trunk is not too large I use an ax, sharpening it on a steel whenever it begins losing its bite. The sharper the cutting edge, the better and easier is the work. Much

FELLING

like the chain saw, I put as little effort as possible into ax work, letting the weight and sharpness of the ax head do the work. On a thick trunk I combine ax and saw, starting off with the hand tool, finishing the deep part of the cut with the saw.

In any event, the notch should extend about a third of the way into the trunk. The bottom surface of the cut as I cut it is flat and horizontal. The top surface goes down and back in. Next step is to saw in from the back, aiming the cut in so it ends up a bit above the bottom level of the notch. Often a tree will settle back on

To fell a tree, notch it on the side facing the direction of fall,

then cut down into the notch from the opposite side.

the cutting bar, stalling the chain, often grabbing the machine as if for keeps. To fix that problem you must pound in cutting wedges that should be made of something softer than the chain so an accidental cut into the wedge won't harm the chain. You can buy wedges made of plastic or soft metals. I chop my own on the spot and if they don't pry the trunk up enough I'll use the steel splitting wedge and be really careful with the saw.

Once the tree begins to fall, ever so slowly at first, I head off, and fast. If everything was done correctly the tree will fall just where it was supposed to, and since the back cut never gets all the way to the notch before the tree begins to fall, the wood left uncut acts as a hinge, keeping the butt of the trunk on the stump till the last second. Then, because of the shape of the cut, the hinge will snap just as the tree hits the ground. That is, if all goes perfectly. Often a tree doesn't know what is expected of it and it will twist in falling, the butt may slide off the stump, or a major limb will catch on another tree and it will pivot. The variations are many and all can kill you. So I get well away, fast.

BUCKING   Next chore is bucking—cutting and splitting the tree. It will be resting on the limbs that grew on the fall side, and often its position is highly tenuous, with a weight imbalance or a tightly sprung branch just waiting to drop the trunk or flip it to one side or another. It pays to check out the supporting limbs, push the tree or even winch the trunk to settle it. Next I start in at the tips of non-supporting limbs and branches, cutting each into a series of foot-

A tree ready for bucking

long stove lengths, throwing the thin stuff into a pile for shredding, burning, bundling into faggots, or just to be left to house a family of cottontails come spring.

     The final task is to cut up the trunk and remaining limbs, which still hold the main part of the tree well off the ground. I'll go in at the former top, still cutting foot-long pieces. When I come to a supporting limb, I cut it with the ax. As each support is removed the trunk will drop, roll or twist, and it is safer to be at ax-length from the trunk than up close with a roaring chain saw.

     Trunks of small trees are a snap to cut. If the trunk needs moving, usually to be raised up on several lengths so the saw will be well clear of the ground, I use a cant hook or peavey, a stout pole with a movable hook on one end that bites into the wood, giving you great leverage. The larger trees are more of a task. Big hollow maples I will cut most of the way through at twelve-inch

SPLITTING LOGS

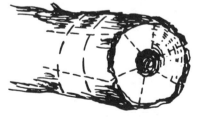

Pattern of cuts in a hollow log

intervals, then go along with the maul and splitting wedge, knocking out chunks of a manageable size. Larger trunks that are solid through, I cut in twelve-inch-thick rounds. All these chunks are hauled in and stored in the barn or on the front porch to be split as needed during the winter. On a really cold day, wood, particularly if still slightly green, becomes brittle and easy to split.

And while trying to split every log into stove wood at one time would be too big a job, splitting each day's supply offers some welcome exercise on winter evenings.

Tanning Hides

A strictly snow-time activity is tanning leather and furs. It's an old saying that you have to work a lot of grease into a properly cured skin—elbow grease. And I'll admit that more than once I've lacked the time or energy to tan a hide or pelt and have sent it off to a commercial tannery. They do a fine job, charge very little, considering, and offer such niceties as reducing a goat or deer skin to uniform thickness (skiving) or dying a sheepskin rug. However, there are some things a tannery can't do and some I wouldn't let them try. First is tanning pig skin.

FLESHING   After a pig or hog is slaughtered in late fall, I tack its hide with the inner, fat-covered side facing out on one wall of the barn.

The very first step in tanning hides is to remove the skin membrane from the hide. Here I am using a fleshing knife to do just that. Next I'll place the hide on a shaping board for drying and storage until I have enough hides to warrant tanning.

All early winter the chickadees, nuthatches, and woodpeckers go at it, receiving their needed fat and protein in turn for doing the job of "fleshing" for us. By midwinter the entire hide has been picked clean of fat and membranes, exposing the tough inner layer that will tan into leather. Without the feathered help I'd have to spend a lot of time and energy working off the flesh with the tanning tools. Though most tanning authorities advise against letting raw hides freeze, I think the alternate freezes and thaws of early winter tend to soften the skin fibers, making tanning easier and faster.

Once the birds have done their work I take the hide down; it will have dried considerably and is fairly stiff. But usually I can roll it up enough that it will go into the cold spring in the cellar. There it stays, in water that is not much above freezing, for a week or two, long enough to soak up well and for the hair and outer surface of the hide to soften and scrape off readily. Then I dress warmly, don rubber boots and gloves, and secure the hide over a length of plastic soil pipe by pressing it between my midsection and the pipe's end. With a square and a curved scraper and a hacksaw, I proceed to work both sides down to the real leather. The hacksaw blade is used to shred any membranes still adhering to the inside, the scrapers to work it free. The outside hair and "grain," or outer skin layers, are worked off with scrapers only.

DOWN TO LEATHER

With some hides a simple water soaking isn't enough to loosen hair and grain. These are put into a barrel filled with a mixture of wood ash and our naturally soft water. In a few weeks time the weak lye from the ashes does the job. The hides are then delimed by being soaked in half a barrel of water with a quart of vinegar added.

Fleshing tool

The Actual Tanning

The final step in changing a hide to leather or a fur is to remove everything in the skin but the tough inner fibers by chemical or mechanical means. If this glue or gluten is not removed completely the hide will decay. Commercial tanneries have a large number of fast tanning processes, each suited to a different kind of skin or end use. We use several old-fashioned procedures that take a lot of time and effort, but few dangerous chemicals.

The original tanning process literally turns a hide the color tan, from the tannic acid in tree bark or other vegetable source. There is a large stand of red sumac in one of our fields; each fall I rake up several feed sacks full of the leaves. (Sumac also provides the wood for our maple sugaring spiles and the fruit for a vitamin-rich jelly—quite a useful little tree.) Another good source of tannic acid is the leaves and bitter acorns from any species of oak

TANNING CHEMISTRY

but the white oak. These we also collect from several big, super-acorn-producers in the nearby woods.

Leaves and acorns are run through the shredder at least twice to reduce them to a fine grit. Then I make up two batches of tanning fluid, each consisting of twenty pounds of vegetable matter, ten gallons of water, and a quart of white vinegar. The mixture is boiled for about a quarter-hour, let cool, then the hides are put in. The tanning vat, well covered, is kept in a handy corner of the upstairs and the skins are stirred around in the liquid whenever we pass by. After two weeks in the vat, I remove the lid. In the dry house the liquid will evaporate and each week I top it up with fresh fluid.

After two months of tanning the hides are removed from the vat, twenty more pounds of ground leaves and acorns are stirred in, and the hides go back in. In another two months I look the hides over and cut a bit into one edge of the neck area. Color will be soaking in from each side. Some small skins will be finished by this time, while thicker ones will need further tanning. Every two months till the color is all the way in, a new batch of leaves and acorns are stirred in and the water is topped up to keep the hides covered.

FINISHING LEATHER

When the tanning process is finished, taking six or eight months with some hides, the leather is removed and let soak in the cellar spring-house well for another few months to let all the acid soak out. Finally I remove it, go over it with the scraper, then oil and soften it. The hide is let partially dry, then I rub neat's-foot oil into both sides, work it in with a scraper and by running the hide back and forth over the edge of a smooth board. The process takes no more than a few minutes with a rabbit skin, much longer with a large one, as the process must be repeated up to a half-dozen times, the skin never being allowed to dry out completely.

The resulting leather is smooth and supple, the color a rich dark brown or black if sumac leaves are used in any quantity. About the only disadvantages, I guess, are the months the skins must spend in the tanning solution and the bother of gathering up a hundred pounds of leaves and acorns. For us though, that is just a good excuse to spend more time in the woods.

Another time-consuming leather-producing method is to make buckskin. Any hide will do, though deer and goat are all we have ever used. The skins are first fleshed and dehaired, then fat or oil is worked into the skin manually, and finally the hide is smoked. The American Indians used the brain of a deer for a fat source. In some tribes the women worked it in by chewing; others rubbed the hide over stakes. We use laundry soap and neat's-foot oil. First the hide is washed well, then, never letting the hide dry out, we work the oil in by running the hide repeatedly over the edges of a sawhorse. After the skin fibers are broken down, the decayable material is worked out and oil worked in by the hours of hand labor. Finally the hide is smoked. We drape the skins on a line strung above the maple sugaring fire. So long as the weather isn't too windy the hides are smoked to a mellow tan by the time the maple supply is in. Then the skin is washed well again in warm water and soap (not detergent), worked over again as it dries, and put in a very special place. Making buckskin is a lot of work and the sort of thing most people would likely try just once. I doubt that we'll make any more till Sam gets old enough to want a real buckskin jacket—and big enough to do most of the work on it.

Furs

If you don't mind dark furs, the sumac leaf and acorn tanning process will do for every skin. On white rabbit skins for little girls' coats and all fancy furs, though, we want to retain the original color. So another tanning process is needed.

TANNING FOR FURS

Hides for furs are fleshed as are any others; the more valuable the hide, the more careful the fleshing. Then they are washed and soaked in warm, soapy water till completely flexible. The tanning mixture is made up of one part saltpeter (potassium nitrate), two parts of alum (potassium aluminum sulfate or potash alum), and three parts of fine sawdust with just enough water to make a paste. Any druggist can get the chemicals for you. The mixture is spread on the skin side of the pelt, an eighth of an inch thick for a small skin, a quarter-inch for a sheep skin or other large hide. The skin is then rolled up tight and kept in the cool of the cellar for a week or two. Basically, the alum soaks in and puckers the skin around the fur while the saltpeter dissolves the matter which would otherwise decay.

Next step is to soak the skin in water, then scrape off all the tanning mix. After it is let dry about halfway, I rinse the skin well in one of the nonflammable dry cleaning solutions obtainable at most hardware stores. This is to remove fat and natural oils which can put out a fairly ripe odor in time. The skin is soaked in water again, and then a light application of neat's-foot oil is

FINISHING FURS

worked in and the skin softened by repeated runnings over the edge of a board or a table.

Leather and furs tanned in these old-fashioned, time- and labor-consuming ways are as durable as if done by a commercial factory, though they lack the even thickness of hide and fluffy gloss of fur you'd find in a $10,000 mink coat. But what we'd have to do to dredge up that kind of cash for that sort of purpose would be a whole lot less pleasant than the hours I spend slogging around in the cellar or roaming the woods to tan hides that come essentially free for the taking.

Holidays on the Homestead

Newly-tanned hides figure prominently in our preparations for Christmas, since we make all gifts for family and friends and much of what we give one another. A favorite with many people is slipper-sox; Louise knits and blocks a pair of warm mid-calf-length sox of the proper size and I cut and stitch on a pair of soles from one or another hide. It is best to have an outline of the recipient's foot to make the soles. I cut a piece of leather an **HOMEMADE** inch wider all around than the foot, cut the extra into flaps at half-**GIFTS** inch intervals, and punch each flap in two places. Then I put a piece

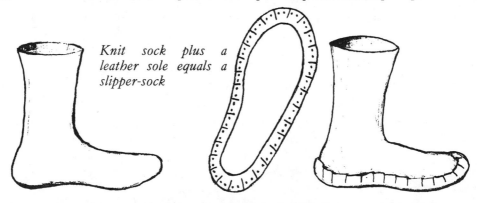

Knit sock plus a leather sole equals a slipper-sock

of heavy cardboard into the sock as if it were a foot, and sew the soles on with an awl and waxed linen thread.

Another favorite that no eight-year-old boy cousin should be without is a 'coon- or skunk-skin cap. First we scout around for a cooking bowl of about the size of the boy's head. Louise makes a felt beanie from four triangles of heavy felt, leaving one seam unsewn. She stitches a length of elastic ribbon around the bottom, leaving a few inches of each end dangling from the edges of the unsewn seam. It will be up to Mom to sew the elastic for a good fit and to adjust it as the boy grows.

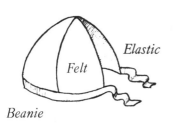

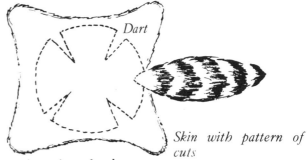

Elastic

Felt

Beanie

Dart

Skin with pattern of cuts

Then I fit the skin to the cap by cutting triangular darts, usually two in front, two in back. Darts are sewed closed, then the cap is put back on the bowl, the elastic tied tight, and I staple the hide to the elastic with a paper stapler. The final step is to trim the hide and stitch it to the elastic with the awl, then pull the staples. A fine gift for the budding Dan'l Boones on our Christmas list.

Pigskin, the thickest part along the back, makes grand summer moccasins since the skin "breathes" better than most. It also stretches to fit its owner's foot so that the preliminary fit needn't be precise. Before cutting into a hide I prefer to make a model of the pattern from heavy old cloth, cut and trimmed to size on the person's foot. Too much hard work goes into tanning to waste the leather on a shoe that is completely the wrong size.

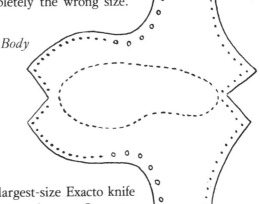

Body

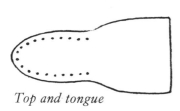

Top and tongue

The pattern is cut out using a largest-size Exacto knife with its razor-sharp blades. Where seams are to be sewn I cut at a forty-five degree angle so the lips of the seam will bevel together in a fine joint. Holes, just punctures for sewing where the skin is thickest and one-eighth-inch holes for laces, are put in the main shoe body with an adjustable leather punch. Seams are closed and the top and tongue are sewn on with the awl. As the pattern will indicate, the result looks none too stylish. But wearing them is the next thing to going barefoot and with a little care they will last half a lifetime.

Skins and furs go into quite a variety of other gifts. Prime winter rabbit fur makes cute collars and cuffs for little girls' outer clothing that Louise sews or knits. Thongs from woodchuck hide or scraps of other skins go on the hanging planters that make good gifts and which sell well during the pre-Christmas buying period. Winters when I run a trapline for mink, weasel, or beaver, Louise has some pretty fancy fur to use in trimming knitted caps and mittens. But since the arrival of little Martha I think I'll start saving the fancy furs, as well as the raccoon, muskrat, skunk, and fox furs we pick up from time to time. There must be a book somewhere that tells how furriers make a really fine garment; and some day our little gal will ride to school in a custom-caught, tanned, tailored, and stitched mixed wild fur coat. Well, maybe she will.

Of course clothing and pottery aren't the only Christmas gifts that our homestead can produce. When the packages go out to distant friends and relatives, many contain an assortment of glistening jars and bottles. Maple syrup or sugar, honey capped with beeswax, pickles and relishes, cheeses, smoked meats, dried fruits and fruit leathers, and other products of the land and our labor are appreciated by many people a great deal more than anything we could purchase.

Throwing pots.

CHRISTMAS

I think the one place where the difference between our old life and the new is most important is in celebration of Christmas. Except for the lights and tinsel on the tree, holdovers from the old days, our counterparts of two hundred years ago would be right at home. We make it a long holiday, just as we used to; as the children grow older we will have an increasingly formal observance of all twelve days of Christmas. So long as they are tinies, though, the holiday will remain mostly presents, Santa, lots of special food, and the tree.

Trimming the tree on our place starts with a gathering of friends and neighbors. Once the adults are prepared for the cold by a bit of mulled wine, we all bundle up and head off into the woods to ''find'' the tree, which often as not has been spotted on an earlier goat walk or hunting trek. I chop the tree down and after we drag it home, set it up, and treat our cold feet with more mulled wine (hot apple cider for the kids), we set to decorating. The children are hustled into the kitchen where the little ones are given bowls of colored icing to decorate the gingerbread cookies Louise has made earlier in the shape of snowmen, stars, and such. A few cookies usually manage to survive to go on the tree.

The oldest child is given the chore of popping the corn, about three times as much as we figure will be needed for the tree. When it's done we all sit around stringing chains of popcorn and cranberries. The ropes go on as completed, followed by whatever cookies have made it through the decorating session. Next we paint the globe ornaments, made the week before by wrapping strips of paper soaked in wheat paste around small, inflated balloons. Painted in whatever bright pattern appeals to the designer, perhaps sprinkled with metallic glitter, they are if anything more colorful than the glass bulbs we used to have, and they don't break. The final touch, our aged tinsel, is put on by Louise and we go in to dinner. On the last day of Christmas, the tree will be stripped of everything but the popcorn and cranberry chains. Then it will be tied to a post beside the front porch bird feeder. The birds will eat the corn and berries and the shelter provided by the branches will attract the shyer species that would otherwise fear to come that close to humans.

Christmas.

The Christmas feasts that Louise serves up on tree-trimming day and Christmas itself will hopefully feature wild turkeys, once the state Fish and Game people decide the recently introduced flocks are strong enough to withstand hunting pressure. For now, though, the main courses will be ham, fresh or home-cured, a leg of venison or chevon, or the biggest capon of the season. The cold cellar contributes jellies, preserves, and wine, the potatoes, the apples for pie, and onions and dried sage for the dressing. From the freezers come an assortment of vegetables and the warm storage bins upstairs turn out winter squash and pumpkin. And the Christmas season is the one time we bake bread and rolls, even if the pace of life with two small children, pigs, goats, and all has sent us sneaking to the bakery more often than we'd like. All of it delicious, home produced—and not a drop of chemicals in a single bite.

A New Year, New Plans

With the end of holiday excitement and activity the tree goes out for the birds, the crust of the last wedge of pumpkin pie has become so soggy it becomes a treat for Horrible Pig, and the homestead slides into a new year. Like most everyone, we find the first few days of January a time for reflection and planning for the coming season. We begin gathering up the receipts and records and when the first seed catalogs arrive I find I'm beginning to plan gardens. But these have become fairly routine tasks by now and most January evenings in front of the living room fire are taken up in discussing the major changes we hope to make on the homestead and our life on the land.

It seems that as time goes by, the projects that we undertake are more and more decided for us by national and world events. Without the recent energy crisis, resulting in skyrocketing costs of power, our planning for an independent water- or wind-powered electricity generator would be a lot more iffy than it is. So would plans for a log splitter, a wood-burning kiln to replace Louise's electric model, and a good deal more. It seems that the pace and severity of change keeps picking up with every passing month, this crisis or that shortage appearing suddenly to make dramatic change in all our lives. With each big change, of course, a homesteader's plans and priorities are affected, though not so drastically as if we were still townspeople. So, after the greenhouse is completed, I really can't say which project will be the next to be carried out. It depends on which will appear to make the most valuable contribution to the ever-expanding partnership between us and the land.

OUR BASIC GOALS   Our basic goal is to become as totally self-sufficient as we can, as free from the money economy and the increasingly severe difficulties of living in a world with too many people and too few resources. Our planning envisions a future—a scenario, the futurists call it—where the United States will have run out of just about every resource but the most valuable, a proportion of the world's arable land that far exceeds our proportion of the world population. Louise and I agree with those economists and natural scientists who predict that within the next few decades the United States will become one huge farm producing (probably from improved soybeans) the protein to feed a hungry world of perhaps twice today's human population.

This means that man's most elemental need, food, will increase in economic value to heights the world has never even imagined; perhaps a bushel of wheat will buy a barrel of crude oil or a new wool suit of clothes. In such a world small-scale farming would again be a viable way of life for millions of American families, and we want to be one of them. Even if this scenario is dead wrong, we want to be able to produce most of our needs here on the homestead; no matter what the future holds for the world, we are convinced that the capacity for self-sufficiency and independence, even if unexercised, is the surest guarantee that life will continue to be good on our little corner of Creation. So, though future plans and projects are pretty mundane in themselves, each is part of a still-evolving plan to establish a self-contained and environmentally sound "homestead of the future."

FOOD SECURITY

So much for generalities. The first item on the agenda is a very down-to-earth building, a greenhouse. We've carried lettuce through most of the winter in the cold frame, but salads get pretty skimpy by January. We want to have a complete winter salad garden with lettuce, cucumbers, and tomatoes all winter long. To date I've picked up some big fluorescent lights, a good neighbor has supplied a pile of old storm sash, and the plans are drawn up. Next winter we'll have it cranked up, insulated with sheets of plastic over glass and heated through a cellar window. I'm sure there will be some mistakes, but with experience we should never again have to put up with lettuce trucked all the way from California or a tomato that was turned red (not ripened) with chemicals.

ENERGY SELF-SUFFICIENCY

On the home-heat front I plan to build a combination log-splitter and cordwood saw that will speed the wood-garnering process. The saw half of the unit will be similar to an antique a neighbor uses to cut his own wood, the splitter from a plan from Garden Way Research of Charlotte, Vermont. I doubt that we will completely discard the central heating unless the cost of heating oil or coal simply goes out of sight. But we have more wood stoves in our future. The little stove in the kitchen will go into my study/workroom as soon as we can find a bigger cookstove with a range and water reservoir. The main part of the house will be heated by one of the modern super-efficient wood or coal heaters such as the Ashley. So far we have reduced our original heating oil consumption by half with the two stoves. The coming heating season, plus two more, should see one tankful last through the winter and in time, as we increase the insulation and finish caulking up the old clapboards, and when the kids are old enough to keep their comforters on at night, we can go to all-wood heat.

We are also following closely the developments in heating with sun power. Our winters are too cold, snowy, and overcast to let us rely exclusively on the solar heat units that have been tried in the Southwest and the Washington, D.C. area. But as more work goes into developing more efficient plants and better heat-storage units, we'll begin experimenting. I'd estimate than in a couple of years, once the diapers are turned to dust cloths and milk-spilling and egg-dropping becomes less frequent, we'll try heating our (greatly reduced) washing water supply with a solar heater. Then in time the south roof of the barn may host a covering of solar panels and half the lowest level of the barn will be changed from goat quarters to a hot water storage tank. With sun-heated water flowing through baseboard heaters to provide general home and cellar heat, and with wood stoves to really warm up rooms in use, we will pretty well be able to forget Exxon, Mobil, and the Arab oil sheiks for good.

The Future in Farming

RAISING
SOME INCOME

There is no question that organically grown food is growing in popularity; here's proof if you need it: the cereal makers are all coming out with "natural" cereals containing whole grains and honey. They are no more organic or really natural than the half-candy stuff they used to feature, but an indication that things are changing. And if, as we expect, food prices continue to soar till small farms run by hand labor become economically sound once again, we will be in the forefront. I am already cutting wood from a partly clear acre-sized meadow a few hundred feet back into our woods, and in time we should have about three relatively level, clear acres for field crops. Some will continue to be used for growing grain for our livestock. The balance will go into the best cash-producing crop we can grow organically with the fertilizers available. Concord grapes are one possibility. Everyone loves to make grape jelly and on the rare occasions we've seen the real Concords for sale the price was amazing.

Other possible crops include popcorn. I even dreamed up what to put on the bags: "All natural, old-time popp'n corn. On the cob, in the shuck. Three (count 'em) or more varieties depending on the crop. Grown the organic way, pure, plain and simple.'' We'd grow a half-dozen different varieties of different colors and flavors. Of course the honey operation can be expanded and will be, if the good doctors can cure me of that sting allergy. Blueberries, the delicious little native variety, can be had for the digging. Strawberries will always be a favorite, and I don't see how we could go wrong with an expanded fruit orchard, perhaps featuring the increasingly popular old-fashioned varieties. And there is always the option to get into maple syrup making or expand the goat herd. Any of these would require a considerable investment in time and equipment, though, neither of which we have in excess. So we'll take it slow.

For our own needs, the farming plans are a bit more precise. We want to increase the sweet corn acreage so as to provide for our corn meal needs, and next summer I plan to try several kinds of wheat for flour. Stock beets or mangels and Jerusalem artichokes will also get a trial as animal feeds; each reportedly produces tremendous crops, though I don't know how they will fare in our climate. To improve culture of root crops, I plan to build a raised planting bed. Old bridge timbers will make a frame approximately six by twelve feet in size and a foot and a half high. Into it will go equal proportions of rich composted rabbit droppings and hay, chopped maple leaves, and crushed bottles. (The shredder-grinder does a great job of reducing bottles to sand-sized cullet that contains nary a splinter. It does so well we use it in the town's glass recycling program.)

On the livestock side, Louise wants to develop a self-perpetuating flock of geese both for food and sale. A pigeon loft will go in over the chicken house and this year or next we'll also try raising turkeys. The vets have pretty well eliminated the disease hazards that formerly made turkey raising hard in any event and impossible near a chicken flock. As soon as I can get the fence up, we'll run a young beef steer or two, and we have entertained the idea of trying to raise American bison, the plains buffalo. Now, don't laugh. They are half again more efficient feed converters than cattle, are much hardier, not as inclined to waste energy putting on fat, and the flavor is similar to beef; many consider it even better. Finally, we plan to develop a strain of homestead-sized hogs. Laboratory scientists have bred them down as small as a big guinea pig. And I would a lot rather dress out two 100-pound hogs than one 250- or 300-pounder. Anyone know where to buy a bison or a miniature pig?

''Shredding'' bottles at the recycling center.

An Independent Power Source

We don't think there is much mystery about the future source of home and farm power for needs other than heat. Petroleum is on its way out; we all know it is going to run out some time. And more and more people are coming to realize that burning it up for transportation and warmth is increasingly foolish. It is probably the world's best industrial raw material, and can be turned into anything from toothbrushes to medicines. In time it simply has to become too costly to waste.

FUTURE POWER SOURCES

Future power will be almost entirely from electricity, in our opinion. The primary source for generating it is the only question still open. Perhaps the atomic establishment will find some way to set up enough nuclear reactors without blowing us all up or poisoning the world with radioactive waste. Maybe the oil companies will be permitted to strip mine Montana for coal and oil shale. Government and industry might even come up with a clean, harmless source such as ocean tides or water temperature variations, wind or solar power, but I don't have much faith that it will happen in our lifetime.

Whatever the source, the result is going to be electricity produced at huge power plants, likely at huge cost. We want to be independent of the whole thing if possible, but unless we go completely savage we can't ever be completely independent of a society that runs everything on electricity. This likely means that in time all our power equipment, autos, and all of home and farm will hum along on electric motors. So we want to have an independent electric generating capability.

WIND AND WATER

With present technology and the costs of the foreign-made propellors and generators, wind power just isn't feasible on our place. We are at the bottom of a hill, surrounded by high country and tall trees, and though the winds howl much of the fall and winter, they are neither strong nor constant enough at other times to supply reliable electric power year-round. We do have a small stream that runs through the old silted-up pond up beyond the back garden wall, but like our well, it slows down in dry autumn seasons. However, that is when the winds are best. So our power plant, if it comes to pass at all, will have to combine water and wind.

First step will be to bulldoze the pond out, enlarging and

deepening it and putting in an earth dam with a concrete spillway. The water wheel will go at the bottom of the spillway and connect through as simple a set of shafts and gears as possible to as big a generator as the water supply will keep going. For the dry and windy months, we'll either have to rely on a wind-powered generator or perhaps put in a new well in the upper pasture and install an old-style windmill pump. Either option will require storage batteries, used direct if we go for a wind-powered generator. With a windmill/pump, we would use the pond for a draw-down reservoir such as several commercial utilities are employing on a larger scale. The windmill would keep pumping water at a constant but relatively slow rate. Every two days or so when the pond was full, we would open the sluice and charge up the batteries. With the low electricity needs of late fall, this should easily power us.

The running costs of such a system would be only a few dollars a year. The initial costs are something else though. When we first looked into wind power the costs were prohibitive, at least $5,000 for an imported unit to meet our needs. But several ecologically-oriented U.S. firms are in process of developing much cheaper models, at less than half the cost. Good-sized generators or alternators suitable for water power can be found at reasonable prices from government surplus or aircraft or railroad salvage sources. (A train or big jet is a moving all-electric hotel, and the generators that supply them can do the same for a less mobile homestead.) Digging the pond and building a dam and spillway could be constructed using a small rented bulldozer and cement mixer, and in addition to electricity the pond will supply Sam a backyard swimming hole, Louise a place to skate, and all of us a meal of fried trout just about any time of year.

Meals of fresh fish and clams are the objective of what is perhaps the ultimate in homestead recycling, growing food from the effluent from the septic system. If anyone finds the idea repellent, I can assure them that what comes from the tap in a good many cities is a lot less sanitary and a lot more hazardous to health than the outflow from a modern septic tank. The water is clean and clear, but packed with "sanitized" nutrients that normally just soak into the soil. What we plan to do is put them to work feeding aquatic food sources. To skeptics, all I can say is that the Chinese have been doing it for about four thousand years and it is working for several experimenters in our country, including the New Alchemy Institute on Cape Cod.

FISH FARMING

First step will be to put in a new septic system which we will need to do soon anyway, as ours is well past retirement age. The system we have in mind would include two concrete tanks, one of a thousand or more gallon capacity for primary treatment, and a smaller one to continue to process effluent and to feed the fish pool. From the small tank would run the standard leach field. (See the diagram.)

The tanks will be buried just beside the house so their tops are about a yard deep. Directly on the tank tops and extending another three feet above ground will be the pool, greenhouse-like shed, and food-growing trays. Putting the pool there saves digging another hole and the heat generated by bacteria digesting sewage will reduce the need for winter heat. The heat supply, electricity, and water will come directly in from the cellar.

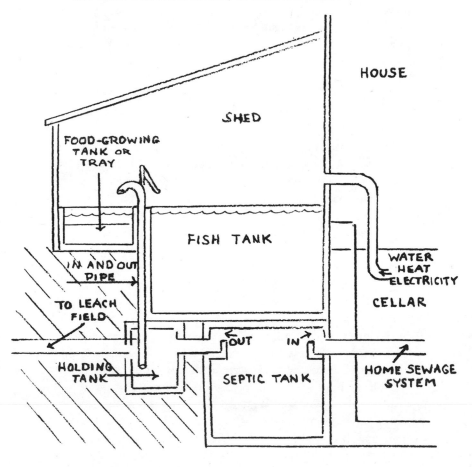

The species of fish and possibly shellfish to be bred is an open question and doubtless will be answered only by trial and error. Some current experimenters are using tropical species often found in aquaria, the mouthbreeders. Chinese grow carp and U.S. fish farms grow both trout and catfish. The question, once any difficulties of keeping water at a desirable temperature are met, is how high on the food chain the breeder wishes to go. Most efficient would be to find species which would live and grow well on green algae which will grow luxuriantly on nothing but effluent and sunlight. Next step up would be to raise carnivorous fish and grow insects or small crustaceans for food. And a considerable proportion of the fish raised, all but the fillets we would consume, could be ground and recycled as feed for the remaining fish. There are a lot of questions needing answers before this idea becomes reality.

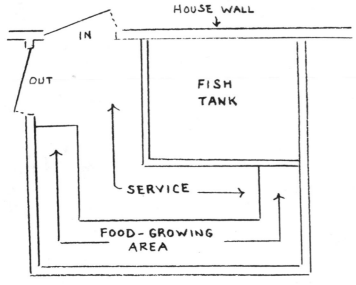

Planning for the Long Term

Well, we've just about run through current grandiose plans. How much will actually get done I can't say. Perhaps changing events or technology will cancel them all and force us to redirect our efforts. However, one thing won't change, our commitment to the land and a life in balance with its rules. Retirement? A homesteader never retires, though I am planting several hundred American black walnut trees to provide the cash to make our last few years a bit easier. The finest of furniture wood, walnuts are getting scarce and a fine log from a conscientiously pruned tree can bring up to $1,000 today. Who knows how much in thirty years?

And finally, we hope to continue our commitment past the end of our lives. If our last will is carried out we will both be buried on our own land, unpickled and in a plain pine box with a wooden monument. In time, we will return completely to the soil that has been so good to us, rather than despoil a piece of earth for all eternity with a bronze casket and cement burial vault. But hopefully that day is a good many years away and we will be able to continue on for decades more of hard work and the deep pleasure of a life in partnership with the land. We hope you will be doing the same.

Postscript

Ironic as it may seem, the inflation and other economic troubles of the mid-1970s promise to hit the "independent" homesteader harder than most city people. So, be forewarned. Louise and I worked long and hard to pare our cash needs to the very bone—and suddenly it is *the bone* that has doubled and tripled in price.

Of course, we aren't much bothered by the rises in food prices that are so serious to townspeople, though it seems that the staples such as rice, cooking oil, pepper and such that we just can't hope to produce ourselves have gone up more than other food items. It is the other basic, essential areas of living cost that we cannot (or have chosen not to) do without that have all jumped in price—petroleum products, electricity, medical insurance. I could go on, but won't.

These price rises affect everyone, but they hit the homesteader the hardest—they have hit us that way. The problem is that we have practically no flexibility to switch our spending around like someone in the conventional money economy. In effect, we are providing *too many* of our own needs to coast through this inflation without a care. For example, since we make our clothes and grow our food I can't delay buying a new suit or Louise can't switch from $1.00 a head fresh broccoli to a 20 cent can of green beans and use the savings to pay the electric company (whose bill charges have risen to about three times what they were just a couple of years ago). We have no choice but to come up with the cash; we can't reduce our consumption really because we've already cut it to the lowest level we care to accept. And that is the case with just about every bill we get; the item is essential to our

INFLATION AND THE HOMESTEADER

way of life, we can not do anything to reduce the amount we use and during the past two years the cash cost has doubled or tripled or worse. And the worst is yet to come.

Now, I'm not complaining. Louise can always turn out a double batch of hanging planters and fortunately there seem to be enough good folks willing to read what I have to say about gardening that I can cut down on bass fishing and spend more time in front of the typewriter. So all I'm saying is, before *you* go a'homesteading, be sure you can do more or less the same thing.

PLAN YOUR INCOME

Before you take off for the country, double check your projections of cash income needs and triple check your source or sources. Compared with a decade ago I'd recommend boosting the priority you assign to getting free of such fixed costs as heating oil or electricity. Our one biggest cost these days is energy of all kinds—gas for cars and tillers, heating oil for the depth of winter when the wood fires leave you singed fore and frozen aft the electricity. If we had it to do over (or if and when we move the homestead—again—to escape "civilization") we would place availability of alternative energy sources high on our list of priorities in a homestead location. Probably even higher than fertility of the soil, which can always be built up organically. But only nature can power an electricity-generating windmill or water wheel, and as I indicated in the last chapter harnessing either wind or water on our place will be a real challenge.

For folks with no hard preference for homestead location I'd strongly suggest checking out Appalachia where hill country land is cheap, water is plentiful, and there are many places where you can dig your own coal from surface seams that are too small for commercial mining. Or you might want to stay close in to a town to reduce travel cost. Might even locate near a genuine city to have public transportation, medical clinics and jobs. Even if it does mean loss of some of the joys of wilderness living (such as the whipporwills which make sleeping through either dawn or dusk on spring evenings an impossibility and the deer which make themselves nuisances in our corn every fall).

So, please be forewarned. And do your income planning carefully and with an eye to an uncertain economic future in America and the world. But for heaven's sake don't let this warning discourage you. I can't think of anyone better prepared to cope with an uncertain future than a well-dug-in homesteader. Oil prices may soar or petroleum get scarce or run out so we either can't afford or can't get the gas to cook our eggs or the electricity to make freezing beans so quick and easy. But we'll always have the eggs. And the beans.

So, once again, join us in the homesteading life.

Index

Page numbers in italics refer to illustrations.